T0181158

Serge Lang

Differential Manifolds

With 17 Illustrations

Springer-Verlag
New York Berlin Heidelberg Tokyo

Serge Lang
Yale University
Department of Mathematics
New Haven, Connecticut 06520
U.S.A.

AMS Classification: 58AXX

Library of Congress Cataloging in Publication Data
Lang, Serge
  Differential manifolds.
  Originally published: Reading, Mass.: Addison-Wesley
Pub. Co., c1972
  Bibliography: p.
  Includes index.
  1. Differentiable manifolds. I. Title.
QA614.3.L35 1985   516.3'6   84-23675

Originally published by Addison-Wesley Publishing Company, Inc., Reading, Massachusetts, copyright 1972.

© 1985 by Springer-Verlag New York Inc.
All rights reserved. No part of this book may be translated or reproduced in any form without written permission from Springer-Verlag, 175 Fifth Avenue, New York, New York 10010, U.S.A.

9 8 7 6 5 4 3 2 1

ISBN 978-0-387-96113-2   ISBN 978-1-4684-0265-0 (eBook)
DOI 10.1007/978-1-4684-0265-0

Foreword

The present volume supersedes my *Introduction to Differentiable Manifolds* written a few years back. I have expanded the book considerably, including things like the Lie derivative, and especially the basic integration theory of differential forms, with Stokes' theorem and its various special formulations in different contexts. The foreword which I wrote in the earlier book is still quite valid and needs only slight extension here.

Between advanced calculus and the three great differential theories (differential topology, differential geometry, ordinary differential equations), there lies a no-man's-land for which there exists no systematic exposition in the literature. It is the purpose of this book to fill the gap.

The three differential theories are by no means independent of each other, but proceed according to their own flavor. In differential topology, one studies for instance homotopy classes of maps and the possibility of finding suitable differentiable maps in them (immersions, embeddings, isomorphisms, etc.). One may also use differentiable structures on topological manifolds to determine the topological structure of the manifold (e.g. à la Smale [26]). In differential geometry, one puts an additional structure on the differentiable manifold (a vector field, a spray, a 2-form, a Riemannian metric, ad lib.) and studies properties connected especially with these objects. Formally, one may say that one studies properties invariant under the group of differentiable automorphisms which preserve the additional structure. In differential equations, one studies vector fields and their integral curves, singular points, stable and unstable manifolds, etc. A certain number of concepts are essential for all three, and are so basic and elementary that it is worth while to collect them together so that more advanced expositions can be given without having to start from the very beginnings. I hope this book will serve this purpose. For a good survey on the current problems of global analysis, cf. Smale [26], and the proceedings of the Berkeley conference [24].

It is possible to lay down *at no extra cost* the foundations (and much more beyond) for manifolds modeled on Banach or Hilbert spaces rather than finite dimensional spaces. In fact, it turns out that the exposition gains considerably from the systematic elimination of the indiscriminate use of

iii

local coordinates $x_1, \ldots, x_n$ and $dx_1, \ldots, dx_n$. These are replaced by what they stand for, namely isomorphisms of open subsets of the manifold on open subsets of Banach spaces (local charts), and a local analysis of the situation which is more powerful and equally easy to use formally. In most cases, the finite dimensional proof extends at once to an invariant infinite dimensional proof. Furthermore, in studying differential forms, one needs to know only the definition of multilinear continuous maps. The orgy of multilinear algebra in standard treatises arises from an unnecessary double dualization and an abusive use of the tensor product.

I don't propose, of course, to do away with local coordinates. They are useful for computations, and are also especially useful when integrating differential forms, because the $dx_1 \cdots dx_n$ corresponds to the $dx_1 \cdots dx_n$ of Lebesgue measure, in oriented charts. Thus we often give the local coordinate formulation for such applications. Most of the literature is still covered by local coordinates, and I therefore hope that the neophyte will thus be helped in getting acquainted with the literature. I also hope to convince the expert that nothing is lost, and much is gained, by expressing one's geometric thoughts without hiding them under an irrelevant formalism.

Infinite dimensional manifolds have been talked about for many years. In recent years, topology has been extremely successful in introducing infinite dimensional topological spaces, and there is every indication that their systematic introduction in the theory of differentiable manifolds will be equally successful. The proper domain for the geodesic part of Morse theory is the loop space, viewed as an infinite dimensional manifold. The reduction to the finite dimensional case is of course a very interesting aspect of the situation, from which one can deduce deep results concerning the finite dimensional manifold itself, but it stops short of a complete analysis of the loop space. (Cf. Bott [3], Milner [18].) This was already mentioned in the first version of the book, and since then, the papers of Palais [23] and Smale [28] appeared, carrying out the program. They determined the appropriate condition in the infinite dimensional case under which this theory works.

In addition, given two finite dimensional manifolds X, Y it is fruitful to give the set of differentiable maps from X to Y an infinite dimensional manifold structure (Eells [8], [9], [10].) In this direction, one would transcend the purely formal translation of finite dimensional results getting essentially new ones, which would in turn affect the finite dimensional case.

Foundations for the geometry of manifolds of mappings are given in Abraham's notes of Smale's lectures [1], and Palais' monograph [23].

The extension of the stable and unstable manifold theorem to the Banach case, already mentioned as a possibility in the earlier version of this book, was proved quite elegantly by Irwin [11], following the idea of Pugh and Robbin for dealing with local flows using the implicit mapping theorem in Banach spaces. I have included the Pugh-Robbin proof, but refer to Irwin's paper for the stable manifold theorem which belongs at the very beginning of the theory of ordinary differential equations. The Pugh-Robbin proof can also be adjusted to hold for vector fields of class H^p (Sobolev spaces), of importance in partial differential equations, as shown by Ebin and Marsden [7].

It is a standard remark that the C^∞-functions on an open subset of a Euclidean space do not form a Banach space. They form a Fréchet space (denumerably many norms instead of one). On the other hand, the implicit function theorem and the local existence theorem for differential equations are not true in the more general case. In order to recover similar results, a much more sophisticated theory is needed, which is only beginning to be developed. (Cf. Nash's paper on Riemannian metrics [21], and subsequent contributions of Schwartz [25] and Moser [19].) In particular, some additional structure must be added (smoothing operators). This goes very much beyond the scope of this book and presents, in fact, an active topic for research.

In writing this book, I have greatly profited from four sources.

First, from Dieudonné's *Foundations of Modern Analysis*. My book is self-contained, starting from Dieudonné's Chapter VIII (Differential Calculus), which is quite elementary, and should be in the curriculum of all advanced undergraduates. The differential calculus in Banach spaces has also been covered in my *Real Analysis*.

Second, from Bourbaki's *Fascicule de résultats* [5] for the foundations of differentiable manifolds. This provides a good guide as to what should be included. I have not followed it entirely, as I have omitted some topics and added others, but on the whole, I found it quite useful. I have put the emphasis on the differentiable point of view, as distinguished from the analytic. However, to offset this a little, I included two analytic applications of Stokes' formula, the Cauchy theorem in several variables, and the residue theorem.

Third, Milnor's notes [15], [16], [17] have proved invaluable. They are of course directed towards differential topology, but of necessity had to cover ad hoc the foundations of differentiable manifolds (or, at least, part of them). In particular, I have used his treatment of the operations on vector bundles (Chapter III, §4) and his elegant exposition of the uniqueness of tubular neighborhoods (Chapter IV, §6 and Chapter VII, §4).

Fourth, I am very much indebted to Palais for collaborating on Chapter IV, and giving me his exposition of sprays (Chapter IV, §3). As he showed me, these can be used (instead of geodesics) to construct tubular neighborhoods. For the relation between sprays and affine connections, the reader is referred to [2]. Palais also showed me how one can recover sprays and geodesics on a Riemannian manifold by making direct use of the fundamental 2-form and the metric (Chapter VII, §6). This is a considerable improvement on past expositions. It is to be hoped that more advanced treatments of differential geometry will be given in the same spirit.

New York   SERGE LANG
August 1971.

Contents

Chapter IX

Stokes' Theorem

Appendix

The Spectral Theorem

CHAPTER I

Differential Calculus

We shall recall briefly the notion of derivative and some of its useful properties. As mentioned in the foreword, Chapter VIII of Dieudonné's book or my *Real Analysis* give a self-contained and complete treatment for Banach spaces. We summarize certain facts concerning their properties as topological vector spaces, and then we summarize differential calculus. *The reader can actually skip this chapter* and start immediately with Chapter II if he is accustomed to thinking about the derivative of a map as a linear transformation. (In the finite dimensional case, when bases have been selected, the entries in the matrix of this transformation are the partial derivatives of the map.) We have repeated the proofs for the more important theorems, for the ease of the reader.

It is convenient to use throughout the language of categories. The notion of category and morphism (whose definitions we recall in §1) is designed to abstract what is common to certain collections of objects and maps between them. For instance, topological vector spaces and continuous linear maps, open subsets of Banach spaces and differentiable maps, differentiable manifolds and differentiable maps, vector bundles and vector bundle maps, topological spaces and continuous maps, sets and just plain maps. In an arbitrary category, maps are called morphisms, and in fact the category of differentiable manifolds is of such importance in this book that from Chapter II on, we use the word morphism synonymously with differentiable map (or p-times differentiable map, to be precise). All other morphisms in other categories will be qualified by a prefix to indicate the category to which they belong.

§1. Categories

A **category** is a collection of objects $\{X, Y, \ldots\}$ such that for two objects X, Y we have a set $\mathrm{Mor}(X, Y)$ and for three objects X, Y, Z a mapping (composition law)

$$\mathrm{Mor}(X, Y) \times \mathrm{Mor}(Y, Z) \to \mathrm{Mor}(X, Z)$$

satisfying the following axioms:

CAT 1. *Two sets* $\text{Mor}(X, Y)$ *and* $\text{Mor}(X', Y')$ *are disjoint unless* $X = X'$ *and* $Y = Y'$, *in which case they are equal.*

CAT 2. *Each* $\text{Mor}(X, X)$ *has an element* id_X *which acts as a left and right identity under the composition law.*

CAT 3. *The composition law is associative.*

The elements of $\text{Mor}(X, Y)$ are called **morphisms**, and we write frequently $f: X \to Y$ for such a morphism. The composition of two morphisms f, g is written fg or $f \circ g$.

A **functor** $\lambda: \mathfrak{A} \to \mathfrak{A}'$ from a category $\mathfrak{A}$ into a category $\mathfrak{A}'$ is a map which associates with each object X in $\mathfrak{A}$ an object $\lambda(X)$ in $\mathfrak{A}'$, and with each morphism $f: X \to Y$ a morphism $\lambda(f): \lambda(X) \to \lambda(Y)$ in $\mathfrak{A}'$ such that, whenever f and g are morphisms in $\mathfrak{A}$ which can be composed, then $\lambda(fg) = \lambda(f)\lambda(g)$ and $\lambda(id_X) = id_{\lambda(X)}$ for all X. This is in fact a covariant functor, and a contravariant functor is defined by reversing the arrows (so that we have $\lambda(f): \lambda(Y) \to \lambda(X)$ and $\lambda(fg) = \lambda(g)\lambda(f)$).

In a similar way, one defines functors of many variables, which may be covariant in some variables and contravariant in others. We shall meet such functors when we discuss multilinear maps, differential forms, etc.

The functors of the same variance from one category $\mathfrak{A}$ to another $\mathfrak{A}'$ form themselves the objects of a category $\text{Fun}(\mathfrak{A}, \mathfrak{A}')$. Its morphisms will sometimes be called **natural transformations** instead of functor morphisms. They are defined as follows. If λ, μ are two functors from $\mathfrak{A}$ to $\mathfrak{A}'$ (say covariant), then a natural transformation $t: \lambda \to \mu$ consists of a collection of morphisms

$$t_X: \lambda(X) \to \mu(X)$$

as X ranges over $\mathfrak{A}$, which makes the following diagram commutative for any morphism $f: X \to Y$ in $\mathfrak{A}$.

$$
\begin{array}{ccc}
\lambda(X) & \xrightarrow{\ t_X\ } & \mu(X) \\
{\scriptstyle \lambda(f)}\downarrow & & \downarrow{\scriptstyle \mu(f)} \\
\lambda(Y) & \xrightarrow{\ t_Y\ } & \mu(Y)
\end{array}
$$

In any category $\mathfrak{A}$, we say that a morphism $f: X \to Y$ is an **isomorphism** if there exists a morphism $g: Y \to X$ such that fg and gf are the identities. For instance, an isomorphism in the category of topological spaces is called a topological isomorphism, or a homeomorphism. In general, we describe the category to which an isomorphism belongs by means of a suitable prefix. In the category of sets, a set-isomorphism is also called a bijection.

If $f: X \to Y$ is a morphism, then a **section** of f is a morphism $g: Y \to X$ such that $f \circ g = id_Y$.

§2. *Topological vector spaces*

The proofs of all statements in this section, including the Hahn-Banach theorem and the closed graph theorem, can be found in my *Real Analysis*.

A **topological vector space** E (over the reals R) is a vector space with a topology such that the operations of addition and scalar multiplication are continuous. It will be convenient to assume also, as part of the definition, that the space is **Hausdorff**, and **locally convex**. By this we mean that every neighborhood of 0 contains an open neighborhood U of 0 such that, if x, y are in U and $0 \leq t \leq 1$, then $tx + (1 - t)y$ also lies in U.

The topological vector spaces form a category, denoted by TVS, if we let the morphisms be the continuous linear maps (by linear we mean throughout R-linear). The set of continuous linear maps of one topological vector space E into F is denoted by $L(\mathbf{E}, \mathbf{F})$. The continuous r-multilinear maps

$$\psi: \mathbf{E} \times \cdots \times \mathbf{E} \to \mathbf{F}$$

of E into F will be denoted by $L^r(\mathbf{E}, \mathbf{F})$. Those which are symmetric (resp. alternating) will be denoted by $L_s^r(\mathbf{E}, \mathbf{F})$ (resp. $L_a^r(\mathbf{E}, \mathbf{F})$). The isomorphisms in the category TVS are called **toplinear** isomorphisms, and we write Lis(E, F) and Laut(E) for the toplinear isomorphisms of E onto F and the toplinear automorphisms of E.

We find it convenient to denote by $L(\mathbf{E})$, $L^r(\mathbf{E})$, $L_s^r(\mathbf{E})$, and $L_a^r(\mathbf{E})$ the continuous linear maps of E into R (resp. the continuous, r-multilinear, symmetric, alternating maps of E into R). Following classical terminology, it is also convenient to call such maps into R **forms** (of the corresponding type). If $\mathbf{E}_1, \ldots, \mathbf{E}_r$ and F are topological vector spaces, then we denote by $L(\mathbf{E}_1, \ldots, \mathbf{E}_r; \mathbf{F})$ the continuous multilinear maps of the product $\mathbf{E}_1 \times \cdots \times \mathbf{E}_r$ into F.

The most important type of topological vector space for us is the **Banachable space** (a TVS which is complete, and whose topology can be defined by a norm). We should say **Banach** space when we want to put the norm into the structure. There are of course many norms which can be used to make a Banachable space into a Banach space, but in practice, one allows the abuse of language which consists in saying Banach space for Banachable space (unless it is absolutely necessary to keep the distinction).

For this book, we assume from now on that all our topological vector spaces are Banach spaces. We shall occasionally make some comments to indicate where it might be possible to generalize certain results to more general spaces. We denote our Banach spaces by $\mathbf{E}, \mathbf{F}, \ldots$.

The next two propositions give two aspects of what is known as the **closed graph theorem**.

Proposition 1. *Every continuous bijective linear map of $\mathbf{E}$ onto $\mathbf{F}$ is a toplinear isomorphism.*

Proposition 2. *If $\mathbf{E}$ is a Banach space, and $\mathbf{F}_1, \mathbf{F}_2$ are two closed subspaces which are complementary (i.e. $\mathbf{E} = \mathbf{F}_1 + \mathbf{F}_2$ and $\mathbf{F}_1 \cap \mathbf{F}_2 = 0$), then the map of $\mathbf{F}_1 \times \mathbf{F}_2$ onto $\mathbf{E}$ given by the sum is a toplinear isomorphism.*

We shall frequently encounter a situation as in Proposition 2, and if $\mathbf{F}$ is a closed subspace of $\mathbf{E}$ such that there exists a closed complement $\mathbf{F}_1$ such that $\mathbf{E}$ is toplinearly isomorphic to the product of $\mathbf{F}$ and $\mathbf{F}_1$ under the natural mapping, then we shall say that $\mathbf{F}$ **splits** in $\mathbf{E}$.

Next, we state a weak form of the Hahn-Banach theorem.

Proposition 3. *Let $\mathbf{E}$ be a topological vector space and $x \neq 0$ an element of $\mathbf{E}$. Then there exists a continuous linear map λ of $\mathbf{E}$ into $\mathbf{R}$ such that $\lambda(x) \neq 0$.*

If $\mathbf{E}$ is a Banach space, one constructs λ by Zorn's lemma, supposing that λ is defined on some subspace, and having a bounded norm. One then extends λ to the subspace generated by one additional element, without increasing the norm. In the general locally convex case, the proof is similar.

In particular, every finite dimensional subspace of $\mathbf{E}$ splits if $\mathbf{E}$ is complete. More trivially, we observe that a finite codimensional closed subspace also splits.

We now come to the problem of putting a topology on $L(\mathbf{E}, \mathbf{F})$. Let $\mathbf{E}, \mathbf{F}$ be Banach spaces, and let

$$A : \mathbf{E} \to \mathbf{F}$$

be a continuous linear map (also called a bounded linear map). We can then define the **norm** of A to be the greatest lower bound of all numbers K such that

$$|Ax| \leq K|x|$$

for all $x \in \mathbf{E}$. This norm makes $L(\mathbf{E}, \mathbf{F})$ into a Banach space.

Remark. For more general topological vector spaces, one would put the topology of uniform convergence on bounded sets.

In a similar way, we define the topology of $L(\mathbf{E}_1, \ldots, \mathbf{E}_r; \mathbf{F})$, which is a Banach space if we define the norm of a multilinear continuous map

$$A : \mathbf{E}_1 \times \cdots \times \mathbf{E}_r \to \mathbf{F}$$

by the greatest lower bound of all numbers K such that

$$|A(x_1, \ldots, x_r)| \leqq K|x_1| \cdots |x_r|.$$

We have:

Proposition 4. *If* $\mathbf{E}_1, \ldots, \mathbf{E}_r, \mathbf{F}$ *are Banach spaces, then the canonical map*

$$L\big(\mathbf{E}_1, L(\mathbf{E}_2, \ldots, L(\mathbf{E}_r, \mathbf{F}), \ldots)\big) \to L'(\mathbf{E}_1, \ldots, \mathbf{E}_r; \mathbf{F})$$

from the repeated continuous linear maps to the continuous multilinear maps is a toplinear isomorphism, which is norm-preserving, i.e. a Banach-isomorphism.

The preceding propositions could be generalized to a wider class of topological vector spaces. The following one exhibits a property peculiar to Banach spaces.

Proposition 5. *Let* $\mathbf{E}, \mathbf{F}$ *be two Banach spaces. Then the set of toplinear isomorphisms* Lis$(\mathbf{E}, \mathbf{F})$ *is open in* $L(\mathbf{E}, \mathbf{F})$.

The proof is in fact quite simple. If Lis$(\mathbf{E}, \mathbf{F})$ is not empty, one is immediately reduced to proving that Laut$(\mathbf{E})$ is open in $L(\mathbf{E}, \mathbf{E})$. We then remark that if $u \in L(\mathbf{E}, \mathbf{E})$, and $|u| < 1$, then the series

$$1 + u + u^2 + \cdots$$

converges. Given any toplinear automorphism w of $\mathbf{F}$, we can find an open neighborhood by translating the open unit ball multiplicatively from 1 to w.

Again in Banach spaces, we have:

Proposition 6. *If* $\mathbf{E}, \mathbf{F}, \mathbf{G}$ *are Banach spaces, then the bilinear maps*

$$L(\mathbf{E}, \mathbf{F}) \times L(\mathbf{F}, \mathbf{G}) \to L(\mathbf{E}, \mathbf{G})$$

$$L(\mathbf{E}, \mathbf{F}) \times \mathbf{E} \to \mathbf{F}$$

obtained by composition of mappings are continuous, and similarly for multilinear maps.

Remark. The preceding proposition is false for more general spaces than Banach spaces, say Fréchet spaces. In that case, one might hope that the following may be true. Let U be open in a Fréchet space and let

$$f \colon U \to L(\mathbf{E}, \mathbf{F})$$

$$g \colon U \to L(\mathbf{F}, \mathbf{G})$$

be continuous. Let γ be the composition of maps. Then $\gamma(f, g)$ is continuous. The same type of question arises later, with differentiable maps instead, and it is of course essential to know the answer to deal with the composition of differentiable maps.

§3. *Derivatives and composition of maps*

A real valued function of a real variable, defined on some neighborhood of 0 is said to be $o(t)$ if

$$\lim_{t \to 0} o(t)/t = 0.$$

Let $\mathbf{E}, \mathbf{F}$ be two topological vector spaces, and φ a mapping of a neighborhood of 0 in $\mathbf{E}$ into $\mathbf{F}$. We say that φ is **tangent to** 0 if, given a neighborhood W of 0 in $\mathbf{F}$, there exists a neighborhood V of 0 in $\mathbf{E}$ such that

$$\varphi(tV) \subset o(t)W$$

for some function $o(t)$. If both $\mathbf{E}, \mathbf{F}$ are normed, then this amounts to the usual condition

$$|\varphi(x)| \leqq |x|\psi(x)$$

with $\lim \psi(x) = 0$ as $|x| \to 0$.

Let $\mathbf{E}, \mathbf{F}$ be two topological vector spaces and U open in $\mathbf{E}$. Let $f \colon U \to \mathbf{F}$ be a continuous map. We shall say that f is **differentiable** at a point $x_0 \in U$ if there exists a continuous linear map λ of $\mathbf{E}$ into $\mathbf{F}$ such that, if we let

$$f(x_0 + y) = f(x_0) + \lambda y + \varphi(y)$$

for small y, then φ is tangent to 0. It then follows trivially that λ is uniquely determined, and we say that it is the **derivative** of f at x_0. We denote the derivative by $Df(x_0)$ or $f'(x_0)$. It is an element of $L(\mathbf{E}, \mathbf{F})$. If f is differentiable at every point of U, then f' is a map

$$f' \colon U \dashrightarrow L(\mathbf{E}, \mathbf{F}).$$

It is easy to verify the chain rule.

Proposition 7. *If $f \colon U \to V$ is differentiable at x_0, if $g \colon V \to W$ is differentiable at $f(x_0)$, then $g \circ f$ is differentiable at x_0, and*

$$(g \circ f)'(x_0) = g'\big(f(x_0)\big) \circ f'(x_0).$$

Proof. We leave it as a simple (and classical) exercise.

The rest of this section is devoted to the statements of the differential calculus. All topological vector spaces are assumed to be Banach spaces (i.e. Banachable). Then $L(\mathbf{E}, \mathbf{F})$ is also a Banach space, if $\mathbf{E}$ and $\mathbf{F}$ are Banach spaces.

Let U be open in $\mathbf{E}$ and let $f\colon U \to \mathbf{F}$ be differentiable at each point of U. If f' is continuous, then we say that f is of class C^1. We define maps of class C^p $(p \geq 1)$ inductively. The p-th derivative $D^p f$ is defined as $D(D^{p-1}f)$ and is itself a map of U into

$$L\big(\mathbf{E}, L(\mathbf{E}, \ldots, L(\mathbf{E}, \mathbf{F}) \cdots)\big)$$

which can be identified with $L^p(\mathbf{E}, \mathbf{F})$ by Proposition 4. A map f is said to be of class C^p if its kth derivative $D^k f$ exists for $1 \leq k \leq p$, and is continuous.

Remark. *Let f be of class C^p, on an open set U containing the origin. Suppose that f is locally homogeneous of degree p near 0, that is*

$$f(tx) = t^p f(x)$$

for all t and x sufficiently small. Then for all sufficiently small x we have

$$f(x) = \frac{1}{p!} D^p f(0) x^{(p)},$$

where $x^{(p)} = (x, x, \ldots, x)$, p times.

This is easily seen by differentiating p times the two expressions for $f(tx)$, and then setting $t = 0$. The differentiation is a trivial application of the chain rule.

Proposition 8. *Let U, V be open in Banach spaces. If $f\colon U \to V$ and $g\colon V \to \mathbf{F}$ are of class C^p, then so is $g \circ f$.*

From Proposition 8, we can view open subsets of Banach spaces as the objects of a category, whose morphisms are the continuous maps of class C^p. These will be called C^p-**morphisms**. We say that f is of class C^∞ if it is of class C^p for all integers $p \geq 1$. From now on, p is an integer ≥ 0 or ∞ (C^0 maps being the continuous maps). In practice, we omit the prefix C^p if the p remains fixed. Thus by **morphism**, throughout the rest of this book, we mean C^p-morphism with $p \leq \infty$. We shall use the word morphism also for C^p-morphisms of manifolds (to be defined in the next chapter), *but morphisms in any other category will always be prefixed so as to indicate the category to which they belong* (for instance bundle morphism, continuous linear morphism, etc.).

Proposition 9. *Let* U *be open in the Banach space* $\mathbf{E}$, *and let* $f: U \to \mathbf{F}$ *be a* C^p*-morphism. Then* $D^p f$ *(viewed as an element of* $L^p(\mathbf{E}, \mathbf{F})$*) is symmetric.*

Proposition 10. *Let* U *be open in* $\mathbf{E}$, *and let* $f_i: U \to \mathbf{F}_i$ $(i = 1, \dots, n)$ *be continuous maps into spaces* $\mathbf{F}_i$. *Let* $f = (f_1, \dots, f_n)$ *be the map of* U *into the product of the* $\mathbf{F}_i$. *Then* f *is of class* C^p *if and only if each* f_i *is of class* C^p, *and in that case*

$$D^p f = (D^p f_1, \dots, D^p f_n).$$

Let U, V be open in spaces $\mathbf{E}_1$, $\mathbf{E}_2$ and let

$$f: U \times V \to \mathbf{F}$$

be a continuous map into a Banach space. We can introduce the notion of partial derivative in the usual manner. If (x, y) is in $U \times V$ and we keep y fixed, then as a function of the first variable, we have the derivative as defined previously. This derivative will be denoted by $D_1 f(x, y)$. Thus

$$D_1 f: U \times V \to L(\mathbf{E}_1, \mathbf{F})$$

is a map of $U \times V$ into $L(\mathbf{E}_1, \mathbf{F})$. We call it the **partial derivative** with respect to the first variable. Similarly, we have $D_2 f$, and we could take n factors instead of 2. The total derivative and the partials are then related as follows.

Proposition 11. *Let* $U_1, \dots, U_n$ *be open in the spaces* $\mathbf{E}_1, \dots, \mathbf{E}_n$ *and let* $f: U_1 \times \cdots \times U_n \to \mathbf{F}$ *be a continuous map. Then* f *is of class* C^p *if and only if each partial derivative* $D_i f: U_1 \times \cdots \times U_n \to L(\mathbf{E}_i, \mathbf{F})$ *exists and is of class* C^{p-1}. *If that is the case, then for* $x = (x_1, \dots, x_n)$ *and*

$$v = (v_1, \dots, v_n) \in \mathbf{E}_1 \times \cdots \times \mathbf{E}_n$$

we have:

$$Df(x) \cdot (v_1, \dots, v_n) = \sum D_i f(x) \cdot v_i.$$

The next four propositions are concerned with continuous linear and multilinear maps.

Proposition 12. *Let* $\mathbf{E}$, $\mathbf{F}$ *be Banach spaces and* $f: \mathbf{E} \to \mathbf{F}$ *a continuous linear map. Then for each* $x \in \mathbf{E}$ *we have*

$$f'(x) = f.$$

Proposition 13. *Let* **E, F, G** *be Banach spaces, and* U *open in* **E**. *Let* $f: U \to \mathbf{F}$ *be of class* C^p *and* $g: \mathbf{F} \to \mathbf{G}$ *continuous and linear. Then* $g \circ f$ *is of class* C^p *and*

$$D^p(g \circ f) = g \circ D^p f.$$

Proposition 14. *If* $\mathbf{E}_1, \dots, \mathbf{E}_r$ *and* **F** *are Banach spaces and*

$$f: \mathbf{E}_1 \times \cdots \times \mathbf{E}_r \to \mathbf{F}$$

a continuous multilinear map, then f *is of class* C^∞, *and its* $(r + 1)$-*st derivative is* 0. *If* $r = 2$, *then* Df *is computed according to the usual rule for derivative of a product (first times the derivative of the second plus derivative of the first times the second).*

Proposition 15. *Let* **E, F** *be Banach spaces which are toplinearly isomorphic. If* $u: \mathbf{E} \to \mathbf{F}$ *is a toplinear isomorphism, we denote its inverse by* u^{-1}. *Then the map*

$$u \mapsto u^{-1}$$

from $\mathrm{Lis}(\mathbf{E}, \mathbf{F})$ *to* $\mathrm{Lis}(\mathbf{F}, \mathbf{E})$ *is a* C^∞-*isomorphism. Its derivative at a point* u_0 *is the linear map of* $L(\mathbf{E}, \mathbf{F})$ *into* $L(\mathbf{F}, \mathbf{E})$ *given by the formula*

$$v \mapsto u_0^{-1} v u_0^{-1}.$$

Finally, we come to some statements which are of use in the theory of vector bundles.

Proposition 16. *Let* U *be open in the Banach space* **E** *and let* **F, G** *be Banach spaces.*

(i) *If* $f: U \to L(\mathbf{E}, \mathbf{F})$ *is a* C^p-*morphism, then the map of* $U \times \mathbf{E}$ *into* **F** *given by*

$$(x, v) \mapsto f(x)v$$

is a morphism.

(ii) *If* $f: U \to L(\mathbf{E}, \mathbf{F})$ *and* $g: U \to L(\mathbf{F}, \mathbf{G})$ *are morphisms, then so is* $\gamma(f, g)$ (γ *being the composition*).

(iii) *If* $f: U \to \mathbf{R}$ *and* $g: U \to L(\mathbf{E}, \mathbf{F})$ *are morphisms, so is* fg (*the value of* fg *at* x *is* $f(x)g(x)$, *ordinary multiplication by scalars*).

(iv) *If* $f, g: U \to L(\mathbf{E}, \mathbf{F})$ *are morphisms, so is* $f + g$.

This proposition concludes our summary of results assumed without proof.

§4. *Integration and Taylor's formula*

Let **E** be a Banach space. Let I denote a real, closed interval, say $a \leq t \leq b$. A **step mapping**

$$f: I \to \mathbf{E}$$

is a mapping such that there exists a finite number of disjoint subintervals $I_1, \ldots, I_n$ covering I such that on each interval I_j, the mapping has constant value, say v_j. We do not require the intervals I_j to be closed. They may be opened, closed, or half-closed.

Given a sequence of mappings f_n from I into **E**, we say that it converges uniformly if, given a neighborhood W of 0 in **E**, there exists an integer n_0 such that, for all $n, m > n_0$ and all $t \in I$, the difference $f_n(t) - f_m(t)$ lies in W. The sequence f_n then converges to a mapping f of I into **E**.

A **ruled** mapping is a uniform limit of step mappings. We leave to the reader the proof that every continuous mapping is ruled.

If f is a step mapping as above, we define its integral

$$\int_a^b f = \int_a^b f(t)\, dt = \sum \mu(I_j) v_j$$

where $\mu(I_j)$ is the length of the interval I_j (its measure in the standard Lebesgue measure). This integral is independent of the choice of intervals I_j on which f is constant.

If f is ruled and $f = \lim f_n$ (lim being the uniform limit), then the sequence

$$\int_a^b f_n$$

converges in **E** to an element of **E** independent of the particular sequence f_n used to approach f uniformly. We denote this limit by

$$\int_a^b f = \int_a^b f(t)\, dt$$

and call it the **integral** of f. The integral is linear in f, and satisfies the usual rules concerning changes of intervals. (If $b < a$ then we define $\int_a^b$ to be minus the integral from b to a.)

As an immediate consequence of the definition, we get:

Proposition 17. *Let $\lambda: \mathbf{E} \to \mathbf{R}$ be a continuous linear map and let $f: I \to \mathbf{E}$ be ruled. Then $\lambda f = \lambda \circ f$ is ruled, and*

$$\lambda \int_a^b f(t)\, dt = \int_a^b \lambda f(t)\, dt.$$

Proof. If f_n is a sequence of step functions converging uniformly to f, then λf_n is ruled and converges uniformly to λf. Our formula follows at once.

Taylor's formula. *Let $\mathbf{E}, \mathbf{F}$ be Banach spaces. Let U be open in $\mathbf{E}$. Let x, y be two points of U such that the segment $x + ty$ lies in U for $0 \leq t \leq 1$. Let*

$$f: U \to \mathbf{F}$$

be a C^p-morphism, and denote by $y^{(p)}$ the "vector" $(y, \ldots, y)$ p times. Then the function $D^p f(x + ty) \cdot y^{(p)}$ is continuous in t, and we have

$$f(x + y) = f(x) + \frac{Df(x)y}{1!} + \cdots + \frac{D^{p-1}f(x)y^{(p-1)}}{(p-1)!}$$

$$+ \int_0^1 \frac{(1 - t)^{p-1}}{(p-1)!} D^p f(x + ty)y^{(p)}\, dt.$$

Proof. By the Hahn-Banach theorem, it suffices to show that both sides give the same thing when we apply a functional λ (continuous linear map into $\mathbf{R}$). This follows at once from Propositions 13 and 17, together with the known result when $\mathbf{F} = \mathbf{R}$. In this case, the proof proceeds by induction on p, and integration by parts, starting from

$$f(x + y) - f(x) = \int_0^1 Df(x + ty)y\, dt.$$

The next two corollaries are known as the **mean value theorem**.

Corollary 1. *Let $\mathbf{E}, \mathbf{F}$ be two Banach spaces, U open in $\mathbf{E}$, and x, z two distinct points of U such that the segment $x + t(z - x)$ $(0 \leq t \leq 1)$ lies in U. Let $f: U \to \mathbf{F}$ be continuous and of class C^1. Then*

$$|f(z) - f(x)| \leq |z - x| \sup |f'(\xi)|,$$

the sup being taken over all ξ in the segment.

Proof. This comes from the usual estimations of the integral. Indeed, for any continuous map $g: I \to \mathbf{F}$ we have the estimate

$$\left| \int_a^b g(t) \, dt \right| \leq K(b - a)$$

if K is a bound for g on I, and $a \leq b$. This estimate is obvious for step functions, and therefore follows at once for continuous functions.

Another version of the mean value theorem is frequently used.

Corollary 2. *Let the hypotheses be as in Corollary 1. Let x_0 be a point on the segment between x and z. Then*

$$|f(z) - f(x) - f'(x_0)(z - x)| \leq |z - x| \sup |f'(\xi) - f'(x_0)|,$$

the sup taken over all ξ on the segment.

Proof. We apply Corollary 1 to the map

$$g(x) = f(x) - f'(x_0)x.$$

Finally, let us make some comments on the estimate of the remainder term in Taylor's formula. We have assumed that $D^p f$ is continuous. Therefore, $D^p f(x + ty)$ can be written

$$D^p f(x + ty) = D^p f(x) + \psi(y, t),$$

where ψ depends on y, t (and x of course), and for fixed x, we have

$$\lim |\psi(y, t)| = 0$$

as $|y| \to 0$. Thus we obtain:

Corollary 3. *Let $\mathbf{E}, \mathbf{F}$ be two Banach spaces, U open in $\mathbf{E}$, and x a point of U. Let $f: U \to \mathbf{F}$ be of class $C^p, p \geq 1$. Then for all y such that the segment $x + ty$ lies in U $(0 \leq t \leq 1)$, we have*

$$f(x + y) = f(x) + \frac{Df(x)y}{1!} + \cdots + \frac{D^p f(x) y^{(p)}}{p!} + \theta(y)$$

with an error term $\theta(y)$ satisfying

$$\lim_{y \to 0} \theta(y)/|y|^p = 0.$$

§5. *The inverse mapping theorem*

The inverse function theorem and the existence theorem for differential equations (of Chapter IV) are based on the next result.

Contraction lemma. Let M be a complete metric space, with distance function d, and let $f: M \to M$ be a mapping of M into itself. Assume that there is a constant K, $0 < K < 1$, such that, for any two points x, y in M, we have

$$d\big(f(x), f(y)\big) \leqq K\, d(x, y).$$

Then f has a unique fixed point (a point such that $f(x) = x$). Given any point x_0 in M, the fixed point is equal to the limit of $f^n(x_0)$ (iteration of f repeated n times) as n tends to infinity.

Proof. This is a trivial exercise in the convergence of the geometric series, which we leave to the reader.

Theorem 1. Let $\mathbf{E}, \mathbf{F}$ be Banach spaces, U an open subset of $\mathbf{E}$, and $f: U \to \mathbf{F}$ a C^p-morphism with $p \geqq 1$. Assume that for some point $x_0 \in U$, the derivative $f'(x_0): \mathbf{E} \to \mathbf{F}$ is a toplinear isomorphism. Then f is a local C^p-isomorphism at x_0.

(By a **local** C^p-**isomorphism** at x_0, we mean that there exists an open neighborhood V of x_0 such that the restriction of f to V establishes a C^p-isomorphism between V and an open subset of $\mathbf{E}$.)

Proof. Since a toplinear isomorphism is a C^∞-isomorphism, we may assume without loss of generality that $\mathbf{E} = \mathbf{F}$ and $f'(x_0)$ is the identity (simply by considering $f'(x_0)^{-1} \circ f$ instead of f). After translations, we may also assume that $x_0 = 0$ and $f(x_0) = 0$.

We let $g(x) = x - f(x)$. Then $g'(x_0) = 0$ and by continuity there exists $r > 0$ such that, if $|x| < 2r$, we have

$$|g'(x)| < \tfrac{1}{2}.$$

From the mean value theorem, we see that $|g(x)| \leqq \tfrac{1}{2}|x|$ and hence g maps the closed ball of radius r, $\bar{B}_r(0)$ into $\bar{B}_{r/2}(0)$.

We contend: Given $y \in \bar{B}_{r/2}(0)$, there exists a unique element $x \in \bar{B}_r(0)$ such that $f(x) = y$. We prove this by considering the map

$$g_y(x) = y + x - f(x).$$

If $|y| \leqq r/2$ and $|x| \leqq r$, then $|g_y(x)| \leqq r$ and hence g_y may be viewed as a mapping of the complete metric space $\bar{B}_r(0)$ into itself. The bound of $\tfrac{1}{2}$

on the derivative together with the mean value theorem shows that g_y is a contracting map, i.e. that

$$|g_y(x_1) - g_y(x_2)| = |g(x_1) - g(x_2)| \leqq \tfrac{1}{2}|x_1 - x_2|$$

for $x_1, x_2 \in \bar{B}_r(0)$. By the contraction lemma, it follows that g_y has a unique fixed point. But the fixed point of g_y is precisely the solution of the equation $f(x) = y$. This proves our contention.

We obtain a local inverse $\varphi = f^{-1}$. This inverse is continuous, because

$$|x_1 - x_2| \leqq |f(x_1) - f(x_2)| + |g(x_1) - g(x_2)|$$

and hence

$$|x_1 - x_2| \leqq 2|f(x_1) - f(x_2)|.$$

Furthermore φ is differentiable in $B_{r/2}(0)$. Indeed, let $y_1 = f(x_1)$ and $y_2 = f(x_2)$ with $y_1, y_2 \in B_{r/2}(0)$ and $x_1, x_2 \in \bar{B}_r(0)$. Then

$$|\varphi(y_1) - \varphi(y_2) - f'(x_2)^{-1}(y_1 - y_2)| = |x_1 - x_2 - f'(x_2)^{-1}(f(x_1) - f(x_2))|.$$

We operate on the expression inside the norm sign with the identity

$$id = f'(x_2)^{-1}f'(x_2).$$

Estimating and using the continuity of f', we see that for some constant A, the preceding expression is bounded by

$$A|f'(x_2)(x_1 - x_2) - f(x_1) + f(x_2)|.$$

From the differentiability of f, we conclude that this expression is $o(x_1 - x_2)$ which is also $o(y_1 - y_2)$ in view of the continuity of φ proved above. This proves that φ is differentiable and also that its derivative is what it should be, namely

$$\varphi'(y) = f'(\varphi(y))^{-1},$$

for $y \in B_{r/2}(0)$. Since the mappings φ, f', "inverse" are continuous, it follows that φ' is continuous and thus that φ is of class C^1. Since taking inverses is C^∞ and f' is C^{p-1}, it follows inductively that φ is C^p, as was to be shown.

Note that this last argument also proves:

Proposition 18. *If $f: U \to V$ is a homeomorphism and is of class C^p with $p \geqq 1$, and if f is a C^1-isomorphism, then f is a C^p-isomorphism.*

In some applications it is necessary to know that if the derivative of a map is close to the identity, then the image of a ball contains a ball of only slightly smaller radius. The precise statement follows. In this book, it will be used only in the proof of the change of variables formula, and therefore may be omitted until the reader needs it.

Lemma. *Let U be open in $\mathbf{E}$, and let $f\colon U \to \mathbf{E}$ be of class C^1. Assume that $f(0) = 0$, $f'(0) = I$. Let $r > 0$ and assume that $\bar{B}_r(0) \subset U$. Let $0 < s < 1$, and assume that*

$$|f'(z) - f'(x)| \leqq s$$

for all $x, z \in \bar{B}_r(0)$. If $y \in \mathbf{E}$ and $|y| \leqq (1 - s)r$, then there exists a unique $x \in \bar{B}_r(0)$ such that $f(x) = y$.

Proof. The map g_y given by $g_y(x) = x - f(x) + y$ is defined for $|x| \leqq r$ and $|y| \leqq (1 - s)r$, and maps $\bar{B}_r(0)$ into itself because, from the estimate

$$|f(x) - x| = |f(x) - f(0) - f'(0)x| \leqq |x| \sup |f'(z) - f'(0)| \leqq sr,$$

we obtain

$$|g_y(x)| \leqq sr + (1 - s)r = r.$$

Furthermore, g_y is a shrinking map because, from the mean value theorem, we get

$$|g_y(x_1) - g_y(x_2)| = |x_1 - x_2 - (f(x_1) - f(x_2))|$$
$$= |x_1 - x_2 - f'(0)(x_1 - x_2) + \delta(x_1, x_2)|$$
$$= |\delta(x_1, x_2)|$$

where

$$|\delta(x_1, x_2)| \leqq |x_1 - x_2| \sup |f'(z) - f'(0)| \leqq s|x_1 - x_2|.$$

Hence g_y has a unique fixed point $x \in \bar{B}_r(0)$ which is such that $f(x) = y$. This proves the lemma.

We shall now prove some useful corollaries, which will be used in dealing with immersions and submersions later. *We assume that morphism means C^p-morphism with $p \geqq 1$.*

Corollary 1. *Let U be an open subset of $\mathbf{E}$, and $f\colon U \to \mathbf{F}_1 \times \mathbf{F}_2$ a morphism of U into a product of Banach spaces. Let $x_0 \in U$, suppose that $f(x_0) = (0, 0)$ and that $f'(x_0)$ induces a toplinear isomorphism of $\mathbf{E}$ and $\mathbf{F}_1 = \mathbf{F}_1 \times 0$. Then there exists a local isomorphism g of $\mathbf{F}_1 \times \mathbf{F}_2$ at $(0, 0)$ such that*

$$g \circ f\colon U \to \mathbf{F}_1 \times \mathbf{F}_2$$

maps an open subset U_1 of U into $\mathbf{F}_1 \times 0$ and induces a local isomorphism of U_1 at x_0 on an open neighborhood of 0 in $\mathbf{F}_1$.

Proof. We may assume without loss of generality that $\mathbf{F}_1 = \mathbf{E}$ (identify by means of $f'(x_0)$) and $x_0 = 0$. We define

$$\varphi : U \times \mathbf{F}_2 \to \mathbf{F}_1 \times \mathbf{F}_2$$

by the formula

$$\varphi(x, y_2) = f(x) + (0, y_2)$$

for $x \in U$ and $y_2 \in \mathbf{F}_2$. Then $\varphi(x, 0) = f(x)$, and

$$\varphi'(0, 0) = f'(0) + (0, id_2).$$

Since $f'(0)$ is assumed to be a toplinear isomorphism onto $\mathbf{F}_1 \times 0$, it follows that $\varphi'(0, 0)$ is also a toplinear isomorphism. Hence by the theorem, it has a local inverse, say g, which obviously satisfies our requirements.

Corollary 1s. *Let $\mathbf{E}, \mathbf{F}$ be Banach spaces, U open in $\mathbf{E}$, and $f : U \to \mathbf{F}$ a C^p-morphism with $p \geq 1$. Let $x_0 \in U$. Suppose that $f(x_0) = 0$ and $f'(x_0)$ gives a toplinear isomorphism of $\mathbf{E}$ on a closed subspace of $\mathbf{F}$ which splits. Then there exists a local isomorphism $g : \mathbf{F} \to \mathbf{F}_1 \times \mathbf{F}_2$ at 0 and an open subset U_1 of U containing x_0 such that the composite map $g \circ f$ induces an isomorphism of U_1 onto an open subset of $\mathbf{F}_1$.*

Considering the splitting assumption, this is a reformulation of Corollary 1.

It is convenient to define the notion of splitting for injections. If $\mathbf{E}, \mathbf{F}$ are topological vector spaces, and $\lambda : \mathbf{E} \to \mathbf{F}$ is a continuous linear map, which is injective, then we shall say that λ **splits** if there exists a toplinear isomorphism $\alpha : \mathbf{F} \to \mathbf{F}_1 \times \mathbf{F}_2$ such that $\alpha \circ \lambda$ induces a toplinear isomorphism of $\mathbf{E}$ onto $\mathbf{F}_1 = \mathbf{F}_1 \times 0$. In our corollary, we could have rephrased our assumption by saying that $f'(x_0)$ is a splitting injection.

For the next corollary, dual to the preceding one, we introduce the notion of a **local projection**. Given a product of two open sets of Banach spaces $V_1 \times V_2$ and a morphism $f : V_1 \times V_2 \to \mathbf{F}$, we say that f is a **projection** (on the first factor) if f can be factored

$$V_1 \times V_2 \to V_1 \to \mathbf{F}$$

into an ordinary projection and an isomorphism of V_1 onto an open subset of $\mathbf{F}$. We say that f is a local projection at (a_1, a_2) if there exists an open neighborhood $U_1 \times U_2$ of (a_1, a_2) such that the restriction of f to this neighborhood is a projection.

Corollary 2. *Let U be an open subset of a product of Banach spaces $E_1 \times E_2$ and (a_1, a_2) a point of U. Let $f: U \to F$ be a morphism into a Banach space, say $f(a_1, a_2) = 0$, and assume that the partial derivative*

$$D_2 f(a_1, a_2) : E_2 \to F$$

is a toplinear isomorphism. Then there exists a local isomorphism h of a product $V_1 \times V_2$ onto an open neighborhood of (a_1, a_2) contained in U such that the composite map

$$V_1 \times V_2 \xrightarrow{\ h\ } U \xrightarrow{\ f\ } F$$

is a projection (on the second factor).

Proof. We may assume $(a_1, a_2) = (0, 0)$ and $E_2 = F$. We define

$$\varphi : E_1 \times E_2 \to E_1 \times E_2$$

by

$$\varphi(x_1, x_2) = \big(x_1, f(x_1, x_2)\big)$$

locally at (a_1, a_2). Then φ' is represented by the matrix

$$\begin{pmatrix} id_1 & 0 \\ D_1 f & D_2 f \end{pmatrix}$$

and is therefore a toplinear isomorphism at (a_1, a_2). By the theorem, it has a local inverse h which clearly satisfies our requirements.

Corollary 2s. *Let U be an open subset of a Banach space E and $f: U \to F$ a morphism into a Banach space F. Let $x_0 \in U$ and assume that $f'(x_0)$ is surjective, and that its kernel splits. Then there exists an open subset U' of U containing x_0 and an isomorphism*

$$h : V_1 \times V_2 \to U'$$

such that the composite map $f \circ h$ is a projection

$$V_1 \times V_2 \to V_1 \to F.$$

Proof. Again this is essentially a reformulation of the corollary, taking into account the splitting assumption.

The implicit mapping theorem. *Let U, V be open sets in Banach spaces E, F respectively, and let*

$$f : U \times V \to G$$

be a C^p mapping. Let $(a, b) \in U \times V$, and assume that

$$D_2 f(a, b): \mathbf{F} \to \mathbf{G}$$

is a toplinear isomorphism. Let $f(a, b) = 0$. Then there exists a continuous map $g: U_0 \to V$ defined on an open neighborhood U_0 of a such that $g(a) = b$ and such that

$$f\big(x, g(x)\big) = 0$$

for all $x \in U_0$. If U_0 is taken to be a sufficiently small ball, then g is uniquely determined, and is also of class C^p.

Proof. Let $\lambda = D_2 f(a, b)$. Replacing f by $\lambda^{-1} \circ f$ we may assume without loss of generality that $D_2 f(a, b)$ is the identity. Consider the map

$$\varphi: U \times V \to \mathbf{E} \times \mathbf{F}$$

given by

$$\varphi(x, y) = \big(x, f(x, y)\big).$$

Then the derivative of φ at (a, b) is immediately computed to be represented by the matrix

$$D\varphi(a, b) = \begin{pmatrix} I_{\mathbf{E}} & O \\ D_1 f(a, b) & D_2 f(a, b) \end{pmatrix} = \begin{pmatrix} I_{\mathbf{E}} & O \\ D_1 f(a, b) & I_{\mathbf{F}} \end{pmatrix}$$

whence φ is locally invertible at (a, b) since the inverse of $D\varphi(a, b)$ exists and is the matrix

$$\begin{pmatrix} I_{\mathbf{E}} & O \\ -D_1 f(a, b) & I_{\mathbf{F}} \end{pmatrix}.$$

We denote the local inverse of φ by ψ. We can write

$$\psi(x, z) = \big(x, h(x, z)\big)$$

where h is some mapping of class C^p. We define

$$g(x) = h(x, 0).$$

Then certainly g is of class C^p and

$$(x, f(x, g(x))) = \varphi\big(x, g(x)\big) = \varphi\big(x, h(x, 0)\big) = \varphi(\psi(x, 0)) = (x, 0).$$

This proves the existence of a C^p map g satisfying our requirements.

Now for the uniqueness, suppose that g_0 is a continuous map defined near a such that $g_0(a) = b$ and $f(x, g_0(x)) = c$ for all x near a. Then $g_0(x)$ is near b for such x, and hence

$$\varphi(x, g_0(x)) = (x, 0).$$

Since φ is invertible near (a, b) it follows that there is a unique point (x, y) near (a, b) such that $\varphi(x, y) = (x, 0)$. Let U_0 be a small ball on which g is defined. If g_0 is also defined on U_0, then the above argument shows that g and g_0 coincide on some smaller neighborhood of a. Let $x \in U_0$ and let $v = x - a$. Consider the set of those numbers t with $0 \leq t \leq 1$ such that $g(a + tv) = g_0(a + tv)$. This set is not empty. Let s be its least **upper** bound. By continuity, we have $g(a + sv) = g_0(a + sv)$. If $s < 1$, we can apply the existence and that part of the uniqueness just proved to show that g and g_0 are in fact equal in a neighborhood of $a + sv$. Hence $s = 1$, and our uniqueness statement is proved, as well as the theorem.

Note. The particular value $f(a, b) = 0$ in the preceding theorem is irrelevant. If $f(a, b) = c$ for some $c \neq 0$, then the above proof goes through replacing 0 by c everywhere.

CHAPTER II

Manifolds

Starting with open subsets of Banach spaces, one can glue them together with C^p-isomorphisms. The result is called a manifold. We begin by giving the formal definition. We then make manifolds into a category, and discuss special types of morphisms. We define the tangent space at each point, and apply the criteria following the inverse function theorem to get a local splitting of a manifold when the tangent space splits at a point.

We shall wait until the next chapter to give a manifold structure to the union of all the tangent spaces.

§1. Atlases, charts, morphisms

Let X be a set. An **atlas of class** C^p $(p \geq 0)$ on X is a collection of pairs (U_i, φ_i) (i ranging in some indexing set), satisfying the following conditions:

AT 1. *Each U_i is a subset of X and the U_i cover X.*

AT 2. *Each φ_i is a bijection of U_i onto an open subset $\varphi_i U_i$ of some Banach space $\mathbf{E}_i$ and for any i, j, $\varphi_i(U_i \cap U_j)$ is open in $\mathbf{E}_i$.*

AT 3. *The map*

$$\varphi_j \varphi_i^{-1} : \varphi_i(U_i \cap U_j) \to \varphi_j(U_i \cap U_j)$$

is a C^p-isomorphism for each pair of indices i, j.

It is a trivial exercise in point set topology to prove that one can give X a topology in a unique way such that each U_i is open, and the φ_i are topological isomorphisms. We see no reason to assume that X is Hausdorff. If we wanted X to be Hausdorff, we would have to place a separation condition on the covering. This plays no role in the formal development in Chapters II and III. It is to be understood, however, that any construction which we perform (like products, tangent bundles, etc.) would yield Hausdorff spaces if we start with Hausdorff spaces.

Each pair (U_i, φ_i) will be called a **chart** of the atlas. If a point x of X lies in U_i, then we say that (U_i, φ_i) is a **chart at** x.

In condition **AT 2**, we did not require that the vector spaces be the same for all indices i, or even that they be toplinearly isomorphic. If they are all equal to the same space $\mathbf{E}$, then we say that the atlas is an **E-atlas**. If two charts (U_i, φ_i) and (U_j, φ_j) are such that U_i and U_j have a non-empty intersection, and if $p \geq 1$, then taking the derivative of $\varphi_j \varphi_i^{-1}$ we see that $\mathbf{E}_i$ and $\mathbf{E}_j$ are toplinearly isomorphic. Furthermore, the set of points $x \in X$ for which there exists a chart (U_i, φ_i) at x such that $\mathbf{E}_i$ is toplinearly isomorphic to a given space $\mathbf{E}$ is both open and closed. Consequently, on each connected component of X, we could assume that we have an **E-atlas** for some fixed $\mathbf{E}$.

Suppose that we are given an open subset U of X and a topological isomorphism $\varphi : U \to U'$ onto an open subset of some Banach space $\mathbf{E}$. We shall say that (U, φ) is **compatible** with the atlas $\{(U_i, \varphi_i)\}$ if each map $\varphi_i \varphi^{-1}$ (defined on a suitable intersection as in **AT 3**) is a C^p-isomorphism. Two atlases are said to be **compatible** if each chart of one is compatible with the other atlas. One verifies immediately that the relation of compatibility bet~een atlases is an equivalence relation. An equivalence class of atlases of class C^p on X is said to define a structure of C^p-**manifold** on X. If all the vector spaces $\mathbf{E}_i$ in some atlas are toplinearly isomorphic, then we can always find an equivalent atlas for which they are all equal, say to the vector space $\mathbf{E}$. We then say that X is an **E-manifold** or that X is **modelled** on $\mathbf{E}$.

If $\mathbf{E} = \mathbf{R}^n$ for some fixed n, then we say that the manifold is n-**dimensional**. In this case, a chart

$$\varphi : U \to \mathbf{R}^n$$

is given by n coordinate functions $\varphi_1, \ldots, \varphi_n$. If P denotes a point of U, these functions are often written

$$x_1(P), \ldots, x_n(P),$$

or simply $x_1, \ldots, x_n$. They are called **local coordinates** on the manifold.

If the integer p (which may also be ∞) is fixed throughout a discussion, we also say that X is a manifold.

The collection of C^p-manifolds will be denoted by Man^p. If we look only at those modelled on spaces in a category $\mathfrak{A}$ then we write $\mathrm{Man}^p(\mathfrak{A})$. Those modelled on a fixed $\mathbf{E}$ will be denoted by $\mathrm{Man}^p(\mathbf{E})$. We shall make these into categories by defining morphisms below.

Let X be a manifold, and U an open subset of X. Then it is possible, in the obvious way, to induce a manifold structure on U, by taking as atlases the intersections

$$(U_i \cap U, \varphi_i \mid (U_i \cap U)).$$

If X is a topological space, covered by open subsets V_j, and if we are given on each V_j a manifold structure such that for each pair j, j' the induced structure on $V_j \cap V_{j'}$ coincides, then it is clear that we can give to X a unique manifold structure inducing the given ones on each V_j.

Example. Let X be the real line, and for each open interval U_i, let φ_i be the function $\varphi_i(t) = t^3$. Then the $\varphi_j \varphi_i^{-1}$ are all equal to the identity, and thus we have defined a C^∞-manifold structure on $\mathbf{R}$!

If X, Y are two manifolds, then one can give the product $X \times Y$ a manifold structure in the obvious way. If $\{(U_i, \varphi_i)\}$ and $\{(V_j, \psi_j)\}$ are atlases for X, Y respectively, then

$$\{(U_i \times V_j, \varphi_i \times \psi_j)\}$$

is an atlas for the product, and the product of compatible atlases gives rise to compatible atlases, so that we do get a well-defined product structure.

Let X, Y be two manifolds. Let $f : X \to Y$ be a map. We shall say that f is a C^p-**morphism** if, given $x \in X$, there exists a chart (U, φ) at x and a chart (V, ψ) at $f(x)$ such that $f(U) \subset V$, and the map

$$\psi \circ f \circ \varphi^{-1} : \varphi U \to \psi V$$

is a C^p-morphism in the sense of Chapter I, §3. One sees then immediately that this same condition holds for any choice of charts (U, φ) at x and (V, ψ) at $f(x)$ such that $f(U) \subset V$.

It is clear that the composite of two C^p-morphisms is itself a C^p-morphism (because it is true for open subsets of vector spaces). The C^p-manifolds and C^p-morphisms form a category. The notion of isomorphism is therefore defined, and we observe that in our example of the real line, the map $t \mapsto t^3$ gives an isomorphism between the funny differentiable structure and the usual one.

If $f : X \to Y$ is a morphism, and (U, φ) is a chart at a point $x \in X$, while (V, ψ) is a chart at $f(x)$, then we shall also denote by

$$f_{V,U} : \varphi U \to \psi V$$

the map $\psi f \varphi^{-1}$.

It is also convenient to have a local terminology. Let U be an open set (of a manifold or a Banach space) containing a point x_0. By a **local isomorphism** at x_0 we mean an isomorphism

$$f: U_1 \to V$$

from some open set U_1 containing x_0 (and contained in U) to an open set V (in some manifold or some Banach space). Thus a local isomorphism is essentially a change of chart, locally near a given point.

§2. Submanifolds, immersions, submersions

Let X be a topological space, and Y a subset of X. We say that Y is **locally closed** in X if every point $y \in Y$ has an open neighborhood U in X such that $Y \cap U$ is closed in U. One verifies easily that a locally closed subset is the intersection of an open set and a closed set. For instance, any open subset of X is locally closed, and any open interval is locally closed in the plane.

Let X be a manifold (of class C^p with $p \geqq 0$). Let Y be a subset of X and assume that for each point $y \in Y$ there exists a chart (V, ψ) at y such that ψ gives an isomorphism of V with a product $V_1 \times V_2$ where V_1 is open in some space $\mathbf{E}_1$ and V_2 is open in some space $\mathbf{E}_2$, and such that

$$\psi(Y \cap V) = V_1 \times a_2$$

for some point $a_2 \in V_2$ (which we could take to be 0). Then it is clear that Y is locally closed in X. Furthermore, the map ψ induces a bijection

$$\psi_1: Y \cap V \to V_1.$$

The collection of pairs $(Y \cap V, \psi_1)$ obtained in the above manner constitutes an atlas for Y, of class C^p. The verification of this assertion, whose formal details we leave to the reader, depends on the following obvious fact.

Lemma 1. *Let U_1, U_2, V_1, V_2 be open subsets of Banach spaces, and $g: U_1 \times U_2 \to V_1 \times V_2$ a C^p-morphism. Let $a_2 \in U_2$ and $b_2 \in V_2$ and assume that g maps $U_1 \times a_2$ into $V_1 \times b_2$. Then the induced map*

$$g_1: U_1 \to V_1$$

is also a morphism.

Indeed, it is obtained as a composite map

$$U_1 \to U_1 \times U_2 \to V_1 \times V_2 \to V_1,$$

the first map being an inclusion and the third a projection.

We have therefore defined a C^p-structure on Y which will be called a **submanifold** of X. This structure satisfies a universal mapping property, which characterises it, namely:

Given any map $f\colon Z \to X$ from a manifold Z into X such that $f(Z)$ is contained in Y. Let $f_Y\colon Z \to Y$ be the induced map. Then f is a morphism if and only if f_Y is a morphism.

The proof of this assertion depends on Lemma 1, and is trivial.

Finally, we note that the inclusion of Y into X is a morphism.

If Y is also a closed subspace of X, then we say that it is a **closed submanifold**.

Suppose that X is finite dimensional of dimension n, and that Y is a submanifold of dimension r. Then from the definition we see that the local product structure in a neighborhood of a point of Y can be expressed in terms of local coordinates as follows. Each point P of Y has an open neighborhood U in X with local coordinates $(x_1, \ldots, x_n)$ such that the points of Y in U are precisely those whose last $n - r$ coordinates are 0, that is, those points having coordinates of type

$$(x_1, \ldots, x_r, 0, \ldots, 0).$$

Let $f\colon Z \to X$ be a morphism, and let $z \in Z$. We shall say that f is an **immersion** at z if there exists an open neighborhood Z_1 of z in Z such that the restriction of f to Z_1 induces an isomorphism of Z_1 onto a submanifold of X. We say that f is an **immersion** if it is an immersion at every point.

Note that there exist injective immersions which are not isomorphisms onto submanifolds, as given by the following example:

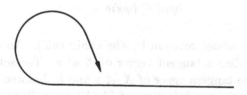

(The arrow means that the line approaches itself without touching.) An immersion which does give an isomorphism onto a submanifold is called an **embedding,** and it is called a **closed embedding** if this submanifold is closed.

A morphism $f\colon X \to Y$ will be called a **submersion** at a point $x \in X$ if there exists a chart (U, φ) at x and a chart (V, ψ) at $f(x)$ such that φ gives an isomorphism of U on a product $U_1 \times U_2$ (U_1 and U_2 open in some Banach spaces), and such that the map

$$\psi f \varphi^{-1} = f_{V,U} \colon U_1 \times U_2 \to V$$

is a projection. One sees then that the image of a submersion is an open subset (a submersion is in fact an open mapping). We say that f is a **submersion** if it is a submersion at every point.

For manifolds modelled on Banach spaces, we have the usual criterion for immersions and submersions in terms of the derivative.

Proposition 1. *Let X, Y be manifolds of class C^p ($p \geq 1$) modelled on Banach spaces. Let $f \colon X \to Y$ be a C^p-morphism. Let $x \in X$. Then:*

(i) *f is an immersion at x if and only if there exists a chart (U, φ) at x and (V, ψ) at $f(x)$ such that $f'_{V,U}(\varphi x)$ is injective and splits.*

(ii) *f is a submersion at x if and only if there exists a chart (U, φ) at x and (V, ψ) at $f(x)$ such that $f'_{V,U}(\varphi x)$ is surjective and its kernel splits.*

Proof. This is an immediate consequence of Corollaries 1 and 2 of the inverse function theorem.

The conditions expressed in (i) and (ii) depend only on the derivative, and if they hold for one choice of charts (U, φ) and (V, ψ) respectively, then they hold for every choice of such charts. It is therefore convenient to introduce a terminology in order to deal with such properties.

Let X be a manifold of class C^p ($p \geq 1$). Let x be a point of X. We consider triples (U, φ, v) where (U, φ) is a chart at x and v is an element of the vector space in which φU lies. We say that two such triples (U, φ, v) and (V, ψ, w) are **equivalent** if the derivative of $\psi \varphi^{-1}$ at φx maps v on w. The formula reads:

$$(\psi \varphi^{-1})'(\varphi x)v = w$$

(obviously an equivalence relation by the chain rule). An equivalence class of such triples is called a **tangent vector** of X at x. The set of such tangent vectors is called the **tangent space** of X at x and is denoted by $T_x(X)$. Each chart (U, φ) determines a bijection of $T_x(X)$ on a Banach space, namely the equivalence class of (U, φ, v) corresponds to the vector v. By means of such a bijection it is possible to transport to $T_x(X)$ the structure of topological vector space given by the chart, and it is immediate that this structure is independent of the chart selected.

If U, V are open in Banach spaces, then to every morphism of class C^p ($p \geq 1$) we can associate its derivative $Df(x)$. If now $f \colon X \to Y$ is a morphism of one manifold into another, and x a point of X, then by means of charts we can interpret the derivative of f on each chart at x as a mapping

$$T_x f \colon T_x(X) \to T_{f(x)}(Y).$$

Indeed, this map $T_x f$ is the unique linear map having the following property. If (U, φ) is a chart at x and (V, ψ) is a chart at $f(x)$ such that $f(U) \subset V$ and $\bar{v}$ is a tangent vector at x represented by v in the chart (U, φ), then

$$T_x f(\bar{v})$$

is the tangent vector at $f(x)$ represented by $Df_{U,V}(x)v$. The representation of $T_x f$ on the spaces of charts can be given in the form of a diagram.

$$
\begin{array}{ccc}
T_x(X) & \longrightarrow & \mathbf{E} \\
T_x f \downarrow & & \downarrow f'_{U,V}(x) \\
T_{f(x)}(Y) & \longrightarrow & \mathbf{F}
\end{array}
$$

The map $T_x f$ is obviously continuous and linear for the structure of topological vector space which we have placed on $T_x(X)$ and $T_{f(x)}(Y)$.

As a matter of notation, we shall sometimes write $f_{*,x}$ instead of $T_x f$.

The operation T satisfies an obvious functorial property, namely, if $f \colon X \to Y$ and $g \colon Y \to Z$ are morphisms, then

$$T_x(g \circ f) = T_{f(x)}(g) \circ T_x(f)$$

$$T_x(id) = id.$$

We may reformulate Proposition 1:

Proposition 2. *Let X, Y be manifolds of class C^p ($p \geq 1$) modelled on Banach spaces. Let $f \colon X \to Y$ be a C^p-morphism. Let $x \in X$. Then:*

(i) *f is an immersion at x if and only if the map $T_x f$ is injective and splits.*

(ii) *f is a submersion at x if and only if the map $T_x f$ is surjective and its kernel splits.*

Note. If X, Y are finite dimensional, then the condition that $T_x f$ splits is superfluous. Every subspace of a finite dimensional vector space splits.

Example. Let $\mathbf{E}$ be a (real) Hilbert space, and let $\langle x, y \rangle \in \mathbf{R}$ be its inner product. Then the square of the norm $f(x) = \langle x, x \rangle$ is obviously of class C^∞. The derivative $f'(x)$ is given by the formula

$$f'(x)y = 2\langle x, y \rangle$$

and for any given $x \neq 0$, it follows that the derivative $f'(x)$ is surjective. Furthermore, its kernel is the orthogonal complement of the subspace generated by x, and hence splits. Consequently the unit sphere in Hilbert space is a submanifold.

If W is a submanifold of a manifold Y of class C^p ($p \geq 1$), then the inclusion

$$i: W \to Y$$

induces a map

$$T_w i: T_w(W) \to T_w(Y)$$

which is in fact an inclusion. From the definition of a submanifold, one sees immediately that the image of $T_w i$ splits. It will be convenient to identify $T_w(W)$ in $T_w(Y)$ if no confusion can result.

A morphism $f: X \to Y$ will be said to be **transversal** over the submanifold W of Y if the following condition is satisfied.

Let $x \in X$ be such that $f(x) \in W$. Let (V, ψ) be a chart at $f(x)$ such that $\psi: V \to V_1 \times V_2$ is an isomorphism on a product, with $\psi(f(x)) = (0, 0)$ and $\psi(W \cap V) = V_1 \times 0$. Then there exists an open neighborhood U of x such that the composite map

$$U \xrightarrow{f} V \xrightarrow{\psi} V_1 \times V_2 \xrightarrow{\mathrm{pr}} V_2$$

is a submersion.

In particular, if f is transversal over W, then $f^{-1}(W)$ is a submanifold of X, because the inverse image of 0 by our local composite map $(\mathrm{pr} \circ \psi \circ f)$ is equal to the inverse image of $W \cap V$ by ψ.

As with immersions and submersions, we have a characterization of transversal maps in terms of tangent spaces.

Proposition 3. *Let X, Y be manifolds of class C^p ($p \geq 1$) modelled on Banach spaces. Let $f: X \to Y$ be a C^p-morphism, and W a submanifold of Y. The map f is transversal over W if and only if for each $x \in X$ such that $f(x)$ lies in W, the composite map*

$$T_x(X) \xrightarrow{T_x f} T_w(Y) \to T_w(Y)/T_w(W)$$

with $w = f(x)$ is surjective and its kernel splits.

Proof. If f is transversal over W, then for each point $x \in X$ such that $f(x)$ lies in W, we choose charts as in the definition, and reduce the question to one of maps of open subsets of Banach spaces. In that case, the conclusion concerning the tangent spaces follows at once from the assumed direct product decompositions. Conversely, assume our condition on the tangent map. The question being local, we can assume that $Y = V_1 \times V_2$ is a product of open sets in Banach spaces such that $W = V_1 \times 0$, and we can also assume that $X = U$ is open in some Banach space, $x = 0$. Then we let $g\colon U \to V_2$ be the map $\pi \circ f$ where π is the projection, and note that our assumption means that $g'(0)$ is surjective and its kernel splits. Furthermore, $g^{-1}(0) = f^{-1}(W)$. We can then use Corollary 2 of the inverse function theorem to conclude the proof.

Remark. In the statement of our proposition, we observe that the surjectivity of the composite map is equivalent to the fact that $T_w(Y)$ is equal to the sum of the image of $T_x f$ and $T_w(W)$, that is

$$T_w(Y) = \operatorname{Im}(T_x f) + \operatorname{Im}(T_x i),$$

where $i\colon W \to Y$ is the inclusion. In the finite dimensional case, the other condition is therefore redundant.

If $\mathbf{E}$ is a Banach space, then the diagonal Δ in $\mathbf{E} \times \mathbf{E}$ is a closed subspace and splits: Either factor $\mathbf{E} \times 0$ or $0 \times \mathbf{E}$ is a closed complement. Consequently, the diagonal is a closed submanifold of $\mathbf{E} \times \mathbf{E}$. If X is any manifold of class C^p, $p \geq 1$, then the diagonal is therefore also a submanifold. (It is closed of course if and only if X is Hausdorff.)

Let $f\colon X \to Z$ and $g\colon Y \to Z$ be two C^p-morphisms, $p \geq 1$. We say that they are **transversal** if the morphism

$$f \times g\colon X \times Y \to Z \times Z$$

is transversal over the diagonal. We remark right away that the surjectivity of the map in Proposition 3 can be expressed in two ways. Given two points $x \in X$ and $y \in Y$ such that $f(x) = g(y) = z$, the condition

$$\operatorname{Im}(T_x f) + \operatorname{Im}(T_y g) = T_z(Z)$$

is equivalent to the condition

$$\operatorname{Im}\big(T_{(x,y)}(f \times g)\big) + T_{(z,z)}(\Delta) = T_{(z,z)}(Z \times Z).$$

Thus in the finite dimensional case, we could take it as definition of transversality.

We use transversality as a sufficient condition under which the fiber product of two morphisms exists. We recall that in any category, the **fiber product** of two morphisms $f\colon X \to Z$ and $g\colon Y \to Z$ over Z consists of an object P and two morphisms

$$g_1\colon P \to X \quad \text{and} \quad g_2\colon P \to Y$$

such that $f \circ g_1 = g \circ g_2$, and satisfying the universal mapping property: Given an object S and two morphisms $u_1 \colon S \to X$ and $u_2 \colon S \to Y$ such that $fu_1 = gu_2$, there exists a unique morphism $u \colon S \to P$ making the following diagram commutative.

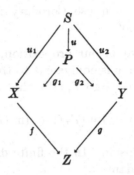

The triple (P, g_1, g_2) is uniquely determined, up to a unique isomorphism (in the obvious sense), and P is also denoted by $X \times_Z Y$.

One can view the fiber product unsymmetrically. Given two morphisms f, g as in the following diagram,

$$
\begin{array}{ccc}
 & & Y \\
 & & \downarrow g \\
X & \xrightarrow{\ f\ } & Z
\end{array}
$$

assume that their fiber product exists, so that we can fill in the diagram:

$$
\begin{array}{ccc}
X \times_Z Y & \longrightarrow & Y \\
{\scriptstyle g_1}\downarrow & & \downarrow{\scriptstyle g} \\
X & \longrightarrow & Z
\end{array}
$$

We say that g_1 is the **pullback** of g by f, and also write it as $f^*(g)$. Similarly, we write $X \times_Z Y$ as $f^*(Y)$.

In our category of manifolds, we shall deal only with cases when the fiber product can be taken to be the set-theoretic fiber product on which a manifold structure has been defined. (The set-theoretic fiber product is the set of pairs of points projecting on the same point.) This determines the fiber product uniquely, and not only up to a unique isomorphism.

Proposition 4. *Let* $f \colon X \to Z$ *and* $g \colon Y \to Z$ *be two* C^p*-morphisms with* $p \geqq 1$. *If they are transversal, then*

$$
(f \times g)^{-1}(\Delta_Z),
$$

together with the natural morphisms into X and Y (obtained from the pro-jections), is a fiber product of f and g over Z.

Proof. Obvious.

To construct a fiber product, it suffices to do it locally. Indeed, let $f: X \to Z$ and $g: Y \to Z$ be two morphisms. Let $\{V_i\}$ be an open covering of Z, and let

$$f_i: f^{-1}(V_i) \to V_i \quad \text{and} \quad g_i: g^{-1}(V_i) \to V_i$$

be the restrictions of f and g to the respective inverse images of V_i. Let $P = (f \times g)^{-1}(\Delta_Z)$. Then P consists of the points (x, y) with $x \in X$ and $y \in Y$ such that $f(x) = g(y)$. We view P as a subspace of $X \times Y$ (i.e. with the topology induced by that of $X \times Y$). Similarly, we construct P_i with f_i and g_i. Then P_i is open in P. The projections on the first and second factors give natural maps of P_i into $f^{-1}(V_i)$ and $g^{-1}(V_i)$, and of P into X and Y.

Proposition 5. *Assume that each P_i admits a manifold structure (com-patible with its topology) such that these maps are morphisms, making P_i into a fiber product of f_i and g_i. Then P, with its natural projections, is a fiber product of f and g.*

To prove the above assertion, we observe that the P_i form a covering of P. Furthermore, the manifold structure on $P_i \cap P_j$ induced by that of P_i or P_j must be the same, because it is the unique fiber product structure over $V_i \cap V_j$, for the maps f_{ij} and g_{ij} (defined on $f^{-1}(V_i \cap V_j)$ and $g^{-1}(V_i \cap V_j)$ respectively). Thus we can give P a manifold structure, in such a way that the two projections into X and Y are morphisms, and make P into a fiber product of f and g.

We shall apply the preceding discussion to vector bundles in the next chapter, and the following local criterion will be useful.

Proposition 6. *Let $f: X \to Z$ be a morphism, and $g: Z \times W \to Z$ be the projection on the first factor. Then f, g have a fiber product, namely the product $X \times W$ together with the morphisms of the following diagram.*

$$
\begin{array}{ccc}
X \times W & \xrightarrow{f \times id} & Z \times W \\
\scriptstyle{\text{pr}_1} \downarrow & & \downarrow \scriptstyle{\text{pr}_1} \\
X & \xrightarrow{f} & Z
\end{array}
$$

§3. Partitions of unity

Let X be a manifold of class C^p. A **function** on X will be a morphism of X into **R**, of class C^p, unless otherwise specified. The C^p functions form a ring $\mathfrak{F}^p(X)$. The **support** of a function f is the closure of the set of points x such that $f(x) \neq 0$.

Let X be a topological space. A covering of X is **locally finite** if every point has a neighborhood which intersects only finitely many elements of the covering. A **refinement** of a covering of X is a second covering, each element of which is contained in an element of the first covering. A topological space is **paracompact** if it is Hausdorff, and every open covering has a locally finite open refinement.

Proposition 7. *If X is a paracompact space, and if $\{U_i\}$ is an open covering, then there exists a locally finite open covering $\{V_i\}$ such that $V_i \subset U_i$ for each i.*

Proof. Let $\{V_k\}$ be a locally finite open refinement of $\{U_i\}$. For each k there is an index $i(k)$ such that $V_k \subset U_{i(k)}$. We let W_i be the union of those V_k such that $i(k) = i$. Then the W_i form a locally finite open covering, because any neighborhood of a point which meets infinitely many W_i must also meet infinitely many V_k.

Proposition 8. *If X is paracompact, then X is normal. If, furthermore, $\{U_i\}$ is a locally finite open covering of X, then there exists a locally finite open covering $\{V_i\}$ such that $\overline{V}_i \subset U_i$.*

Proof. We refer the reader to Bourbaki [6].

Observe that Proposition 7 shows that the insistence that the indexing set of a refinement be a given one can easily be achieved.

A **partition of unity** (of class C^p) on a manifold X consists of an open covering $\{U_i\}$ of X and a system of functions

$$\psi_i: X \to \mathbf{R}$$

satisfying the following conditions.

PU 1. *For all $x \in X$ we have $\psi_i(x) \geqq 0$.*

PU 2. *The support of ψ_i is contained in U_i.*

PU 3. *The covering is locally finite.*

PU 4. *For each point $x \in X$ we have*

$$\sum \psi_i(x) = 1.$$

(The sum is taken over all i, but is in fact finite for any given point x in view of **PU 3**.)

We sometimes say that $\{(U_i, \psi_i)\}$ is a partition of unity.

A manifold X will be said to **admit partitions of unity** if it is paracompact, and if, given a locally finite open covering $\{U_i\}$, there exists a partition of unity $\{\psi_i\}$ such that the support of ψ_i is contained in U_i.

If $\{U_i\}$ is a covering of X, then we say that a covering $\{V_k\}$ is subordinated to $\{U_i\}$ if each V_k is contained in some U_i.

It is desirable to give sufficient conditions on a manifold in order to insure the existence of partitions of unity. There is no difficulty with the topological aspects of this problem. It is known that a metric space is paracompact (cf. Bourbaki [6]), and on a paracompact space, one knows how to construct continuous partitions of unity (loc. cit.). However, in the case of infinite dimensional manifolds, certain difficulties arise to construct differentiable ones, and it is known that a Banach space itself may not admit partitions of unity (say of class C^∞). In the finite dimensional case, the existence will follow from the next theorem.

If **E** is a Banach space, we denote by $B_r(a)$ the open ball of radius r and center a, and by $\bar{B}_r(a)$ the closed ball of radius r and center a. If $a = 0$, then we write B_r and $\bar{B}_r$ respectively. Two open balls (of finite radius) are obviously C^∞-isomorphic. If X is a manifold and (V, φ) is a chart at a point $x \in X$, then we say that (V, φ) (or simply V) is a ball of radius r if φV is a ball of radius r in the Banach space.

Theorem 1. *Let X be a manifold which is locally compact, Hausdorff, and whose topology has a countable base. Given an open covering of X, then there exists an atlas $\{(V_k, \varphi_k)\}$ such that the covering $\{V_k\}$ is locally finite and subordinated to the given covering, such that $\varphi_k V_k$ is the open ball B_3, and such that the open sets $W_k = \varphi_k^{-1}(B_1)$ cover X.*

Proof. Let $U_1, U_2, \ldots$ be a basis for the open sets of X such that each $\bar{U}_1$ is compact. We construct inductively a sequence $A_1, A_2, \ldots$ of compact sets whose union is X, such that A_i is contained in the interior of A_{i+1}. We let $A_1 = \bar{U}_1$. Suppose we have constructed A_i. We let j be the smallest integer such that A_i is contained in $U_1 \cup \cdots \cup U_j$. We let A_{i+1} be the closed and compact set

$$\bar{U}_1 \cup \cdots \cup \bar{U}_j \cup \bar{U}_{i+1}.$$

For each point $x \in X$ we can find an arbitrarily small chart (V_x, φ_x) at x such that $\varphi_x V_x$ is the ball of radius 3 (so that each V_x is contained in some element of U). We let $W_x = \varphi_x^{-1}(B_1)$ be the ball of radius 1 in this chart. We can cover the set

$$A_{i+1} - \text{Int}(A_i)$$

(intuitively the closed annulus) by a finite number of these balls of radius 1, say $W_1, \ldots, W_n$, such that, at the same time, each one of $V_1, \ldots, V_n$ is contained in the open set Int $(A_{i+2}) - A_{i-1}$ (intuitively, the open annulus of the next bigger size). We let $\mathfrak{B}_i$ denote the collection $V_1, \ldots, V_n$ and let $\mathfrak{B}$ be composed of the union of the $\mathfrak{B}_i$. Then $\mathfrak{B}$ is locally finite, and we are done.

Corollary. *Let X be a manifold which is locally compact Hausdorff, and whose topology has a countable base. Then X admits partitions of unity.*

Proof. Let $\{(V_k, \varphi_k)\}$ be as in the theorem, and $W_k = \varphi_k^{-1}(B_1)$. We can find a function ψ_k of class C^p such that $0 \leq \psi_k \leq 1$, such that $\psi_k(x) = 1$ for $x \in W_k$ and $\psi_k(x) = 0$ for $x \notin V_k$. (The proof is recalled below.) We now let

$$\psi = \sum \psi_k$$

(a sum which is finite at each point), and we let $\gamma_k = \psi_k/\psi$. Then $\{(V_k, \gamma_k)\}$ is the desired partition of unity.

We now recall the argument giving the function ψ_k. First, given two real numbers r, s with $0 \leq r < s$, the function defined by

$$\exp\left(\frac{-1}{(t-r)(s-t)}\right)$$

in the open interval $r < t < s$ and 0 outside the interval determines a bell-shaped C^∞-function from $\mathbf{R}$ into $\mathbf{R}$. Its integral from minus infinity to t, divided by the area under the bell yields a function which lies strictly between 0 and 1 in the interval $r < t < s$, is equal to 0 for $t \leq r$ and is equal to 1 for $t \geq s$. (The function is even monotone increasing.)

We can therefore find a real valued function of a real variable, say $\eta(t)$, such that $\eta(t) = 1$ for $|t| < 1$ and $\eta(t) = 0$ for $|t| \geq 1 + \delta$ with small δ, and such that $0 \leq \eta \leq 1$. If $\mathbf{E}$ is a Hilbert space, then $\eta(|x|^2) = \psi(x)$ gives us a function which is equal to 1 on the ball of radius 1 and 0 outside the ball of radius $1 + \delta$. This function can then be transported to the manifold by any given chart whose image is the ball of radius 3.

In a similar way, one would construct a function which is > 0 on a given ball and $= 0$ outside this ball.

Partitions of unity constitute the only known means of gluing together local mappings (into objects having an addition, namely vector bundles, discussed in the next chapter). It is therefore important, in both the Banach and Hilbert cases, to determine conditions under which they exist. In the Banach case, there is the added difficulty that the argument just given to get a local function which is 1 on B_1 and 0 outside B_2 fails if one cannot find a differentiable function of the norm, or of an equivalent norm used to define the Banachable structure. The existence of such functions is not known as this book is written.

Even though it is not known whether Theorem 1 extends to Hilbert manifolds, it is still possible to construct partitions of unity in that case. As Eells pointed out to me, Dieudonné's method of proof showing that separable metric space is paracompact can be applied for that purpose (this is Lemma 1 below), and I am indebted to him for the following exposition.

We need some lemmas. We use the notation ^cA for the complement of a set A.

Let M be a metric space with distance function d. We can then speak of open and closed balls. For instance $\bar{B}_a(x)$ denotes the closed ball of radius a with center x. It consists of all points y with $d(y, x) \leq a$. An open subset V of M will be said to be **scalloped** if there exist open balls $U, U_1, \ldots, U_m$ in M such that

$$V = U \cap {}^c\bar{U}_1 \cap \cdots \cap {}^c\bar{U}_m.$$

A covering $\{V_i\}$ of a subset W of M is said to be locally finite (with respect to W) if every point $x \in W$ has a neighborhood which meets only a finite number of elements of the covering.

Lemma 1. *Let M be a metric space and $\{U_i\}$ $(i = 1, 2, \ldots)$ a countable covering of a subset W by open balls. Then there exists a locally finite open covering $\{V_i\}$ $(i = 1, 2, \ldots)$ of W such that $V_i \subset U_i$ for all i, and such that V_i is scalloped for all i.*

Proof. We define V_i inductively as follows. Each U_i is a ball, say $B_{a_i}(x_i)$. Let $V_1 = U_1$. Having defined V_{i-1}, let

$$r_{1i} = a_1 - \frac{1}{i}, \quad \ldots, \quad r_{i-1,i} = a_{i-1} - \frac{1}{i},$$

and let

$$V_i = U_i \cap {}^c\bar{B}_{r_{1i}}(x_1) \cap \cdots \cap {}^c\bar{B}_{r_{i-1,i}}(x_{i-1}),$$

it being understood that a ball of negative radius is empty. Then each V_i is scalloped, and is contained in U_i. We contend that the V_i cover W. Indeed, let x be an element of W. Let j be the smallest index such that $x \in U_j$. Then $x \in V_j$, for otherwise, x would be in the complement of V_j which is equal to the union of ^cU_j and the balls

$$\bar{B}_{r_{1j}}(x_1) \cup \cdots \cup \bar{B}_{r_{j-1,j}}(x_{j-1}).$$

Hence x would lie in some U_i with $i < j$, contradiction.

There remains to be shown that our covering $\{V_i\}$ is locally finite. Let $x \in W$. Then x lies in some U_n. Let s be a very small number > 0 such that the ball $B_s(x)$ is contained in U_n. Let $t = s/2$. For all i sufficiently large, the ball $B_i(x)$ is contained in $\bar{B}_{a_n - 1/t}(x_n) = \bar{B}_{r_{ni}}(x_n)$ and therefore this ball does not meet V_i. We have found a neighborhood of x which meets only a finite number of members of our covering, which is consequently locally finite (with respect to W).

Lemma 2. *Let U be an open ball in Hilbert space* **E** *and let*

$$V = U \cap {}^c\bar{U}_1 \cap \cdots \cap {}^c\bar{U}_m$$

be a scalloped open subset. Then there exists a C^∞-function $\omega \colon \mathbf{E} \to \mathbf{R}$ such that $\omega(x) > 0$ if $x \in V$ and $\omega(x) = 0$ otherwise.

Proof. For each U_i let $\varphi_i \colon \mathbf{E} \to \mathbf{R}$ be a function such that

$$0 \leq \varphi_i(x) < 1 \qquad \text{if } x \in {}^c\bar{U}_i,$$

$$\varphi_i(x) = 1 \qquad \text{if } x \in \bar{U}_i.$$

Let $\varphi(x)$ be a function such that $\varphi(x) > 0$ on U and $\varphi(x) = 0$ outside U. Let

$$\omega(x) = \varphi(x) \prod \big(1 - \varphi_i(x)\big).$$

Then $\omega(x)$ satisfies our requirements.

Theorem 2. *Let A_1, A_2 be non-void, closed, disjoint subsets of a separable Hilbert space* **E**. *Then there exists a C^∞-function $\psi \colon \mathbf{E} \to \mathbf{R}$ such that $\psi(x) = 0$ if $x \in A_1$ and $\psi(x) = 1$ if $x \in A_2$, and $0 \leq \psi(x) \leq 1$ for all x.*

Proof. By Lindelöf's theorem, we can find a countable collection of open balls $\{U_i\}$ $(i = 1, 2, \ldots)$ covering A_2 and such that each U_i is contained in the complement of A_1. Let W be the union of the U_i. We find a locally finite refinement $\{V_i\}$ as in Lemma 1. Using Lemma 2, we find a function ω_i which is > 0 on V_i and 0 outside V_i. Let $\omega = \sum \omega_i$ (the sum is finite at each point of W). Then $\omega(x) > 0$ if $x \in A_2$, and $\omega(x) = 0$ if $x \in A_1$.

Let U be the open neighborhood of A_2 on which ω is > 0. Then A_2 and ${}^cU$ are disjoint closed sets, and we can apply the above construction to obtain a function $\sigma \colon \mathbf{E} \to \mathbf{R}$ which is > 0 on ${}^cU$ and $= 0$ on A_2. We let $\psi = \omega/(\sigma + \omega)$. Then ψ satisfies our requirements.

Corollary. *Let X be a paracompact manifold of class C^p, modelled on a separable Hilbert space* **E**. *Then X admits partitions of unity (of class C^p).*

Proof. It is trivially verified that an open ball of finite radius in **E** is C^∞-isomorphic to **E**. (We reproduce the formula in Chapter VII.) Given any point $x \in X$, and a neighborhood N of x, we can therefore always find a chart (G, γ) at x such that $\gamma G = \mathbf{E}$, and $G \subset N$. Hence, given an open covering of X, we can find an atlas $\{(G_\alpha, \gamma_\alpha)\}$ subordinated to the given covering, such that $\gamma_\alpha G_\alpha = \mathbf{E}$. By paracompactness, we can find a refinement $\{U_i\}$ of the covering $\{G_\alpha\}$ which is locally finite. Each U_i is contained in some $G_{\alpha(i)}$ and we let φ_i be the restriction of $\gamma_{\alpha(i)}$ to U_i. We now find open refinements $\{V_i\}$ and then $\{W_i\}$ such that

$$\overline{W}_i \subset V_i \subset \overline{V}_i \subset U_i,$$

the bar denoting closure in X. Each $\overline{V}_i$ being closed in X, it follows from our construction that $\varphi_i \overline{V}_i$ is closed in $\mathbf{E}$, and so is $\varphi_i \overline{W}_i$. Using the theorem, and transporting functions on $\mathbf{E}$ to functions on X by means of the φ_i, we can find for each i a C^p-function $\psi_i : X \to \mathbf{R}$ which is 1 on $\overline{W}_i$ and 0 on $X - V_i$. We let $\psi = \sum \psi_i$ and $\theta_i = \psi_i / \psi$. Then the collection $\{\theta_i\}$ is the desired partition of unity.

§4. Manifolds with boundary

Let $\mathbf{E}$ be a Banach space, and $\lambda : \mathbf{E} \to \mathbf{R}$ a continuous linear map into $\mathbf{R}$. (This will also be called a **functional** on $\mathbf{E}$.) We denote by $\mathbf{E}_\lambda^0$ the kernel of λ, and by $\mathbf{E}_\lambda^+$ (resp. $\mathbf{E}_\lambda^-$) the set of points $x \in \mathbf{E}$ such that $\lambda(x) \geq 0$ (resp. $\lambda(x) \leq 0$). We call $\mathbf{E}_\lambda^0$ a **hyperplane** and $\mathbf{E}_\lambda^+$ or $\mathbf{E}_\lambda^-$ a **half plane**.

If μ is another functional and $\mathbf{E}_\lambda^+ = \mathbf{E}_\mu^+$, then there exists a number $c > 0$ such that $\lambda = c\mu$. This is easily proved. Indeed, we see at once that the kernels of λ and μ must be equal. Suppose $\lambda \neq 0$. Let x_0 be such that $\lambda(x_0) > 0$. Then $\mu(x_0) > 0$ also. The functional $\lambda - (\lambda(x_0)/\mu(x_0))\mu$ vanishes on the kernel of λ (or μ) and also on x_0. Therefore it is the 0 functional, and $c = \lambda(x_0)/\mu(x_0)$.

Let $\mathbf{E}, \mathbf{F}$ be Banach spaces, and let $\mathbf{E}_\lambda^+$ and $\mathbf{F}_\mu^+$ be two half planes in $\mathbf{E}$ and $\mathbf{F}$ respectively. Let U, V be two open subsets of these half planes respectively. We shall say that a mapping

$$f : U \to V$$

is a morphism of class C^p if the following condition is satisfied. Given a point $x \in U$, there exists an open neighborhood U_1 of x in $\mathbf{E}$, an open neighborhood V_1 of $f(x)$ in $\mathbf{F}$, and a morphism $f_1 : U_1 \to V_1$ (in the sense of Chapter I) such that the restriction of f_1 to $U_1 \cap U$ is equal to f. (We assume that all morphisms are of class C^p with $p \geq 1$.)

If our half planes are full planes (i.e. equal to the vector spaces themselves), then our present definition is the same as the one used previously.

If we take as objects the open subsets of half planes in Banach spaces, and as morphisms the C^p-morphisms, then we obtain a category. The notion of isomorphism is therefore defined, and the definition of manifold by means of atlases and charts can be used as before. The manifolds of §1 should have been called **manifolds without boundary**, reserving the name of manifold for our new globalized objects. However, in most of this book, we shall deal exclusively with manifolds without boundary for simplicity. The following remarks will give the reader the means of extending any result he wishes (provided it is true) from the case of manifolds without boundaries to the case of manifolds with.

First, concerning the notion of derivative, we have:

Proposition 9. *Let $f: U \to \mathbf{F}$ and $g: U \to \mathbf{F}$ be two morphisms of class C^p ($p \geqq 1$) defined on an open subset U of $\mathbf{E}$. Assume that f and g have the same restriction to $U \cap \mathbf{E}_\lambda^+$ for some half plane $\mathbf{E}_\lambda^+$, and let $x \in U \cap \mathbf{E}_\lambda^+$. Then $f'(x) = g'(x)$.*

Proof. After considering the difference of f and g, we may assume without loss of generality that the restriction of f to $U \cap \mathbf{E}_\lambda^+$ is 0. It is then obvious that $f'(x) = 0$.

Proposition 10. *Let U be open in $\mathbf{E}$. Let μ be a non-zero functional on $\mathbf{F}$ and let $f: U \to \mathbf{F}_\mu^+$ be a morphism of class C^p with $p \geqq 1$. If x is a point of U such that $f(x)$ lies in $\mathbf{F}_\mu^0$ then $f'(x)$ maps $\mathbf{E}$ into $\mathbf{F}_\mu^0$.*

Proof. Without loss of generality, we may assume that $x = 0$ and $f(x) = 0$. Let W be a given neighborhood of 0 in $\mathbf{F}$. Suppose that we can find a small element $v \in \mathbf{E}$ such that $\mu f'(0)v \neq 0$. We can write (for small t):

$$f(tv) = tf'(0)v + o(t)w_t$$

with some element $w_t \in W$. By assumption, $f(tv)$ lies in $\mathbf{F}_\mu^+$. Applying μ we get

$$t\mu f'(0)v + o(t)\mu(w_t) \geqq 0.$$

Dividing by t, this yields

$$\mu f'(0)v \geqq \frac{o(t)}{t} \mu(w_t).$$

Replacing t by $-t$, we get a similar inequality on the other side. Letting t tend to 0 shows that $\mu f'(0)v = 0$, a contradiction.

Let U be open in some half plane $\mathbf{E}_\lambda^+$. We define the **boundary** of U (written ∂U) to be the intersection of U with $\mathbf{E}_\lambda^0$, and the **interior** of U

(written Int (U)) to be the complement of ∂U in U. Then Int (U) is open in $\mathbf{E}$.

It follows at once from our definition of differentiability that a half plane is C^∞-isomorphic with a product

$$\mathbf{E}_\lambda^+ \approx \mathbf{E}_\lambda^0 \times \mathbf{R}^+$$

where $\mathbf{R}^+$ is the set of real numbers $\geqq 0$, whenever $\lambda \neq 0$. The boundary of $\mathbf{E}_\lambda^+$ in that case is $\mathbf{E}_\lambda^0 \times 0$.

Proposition 11. *Let λ be a functional on $\mathbf{E}$ and μ a functional on $\mathbf{F}$. Let U be open in $\mathbf{E}_\lambda^+$ and V open in $\mathbf{F}_\mu^+$ and assume $U \cap \mathbf{E}_\lambda^0$, $V \cap \mathbf{F}_\mu^0$ are not empty. Let $f: U \to V$ be an isomorphism of class C^p ($p \geqq 1$). Then $\lambda \neq 0$ if and only if $\mu \neq 0$. If $\lambda \neq 0$, then f induces a C^p-isomorphism of Int (U) on Int (V) and of ∂U on ∂V.*

Proof. By the functoriality of the derivative, we know that $f'(x)$ is a toplinear isomorphism for each $x \in U$. Our first assertion follows from the preceding proposition. We also see that no interior point of U maps on a boundary point of V and conversely. Thus f induces a bijection of ∂U on ∂V and a bijection of Int (U) on Int (V). Since these interiors are open in their respective spaces, our definition of derivative shows that f induces an isomorphism between them. As for the boundary, it is a submanifold of the full space, and locally, our definition of derivative, together with the product structure, shows that the restriction of f to ∂U must be an isomorphism on ∂V.

This last proposition shows that the boundary is a differentiable invariant, and thus that we can speak of the boundary of a manifold.

We give just two words of warning concerning manifolds with boundary. First, products do not exist in their category. Indeed, to get products, we are forced to define manifolds with **corners**, which would take us too far afield.

Second, in defining immersions or submanifolds, there is a difference in kind when we consider a manifold embedded in a manifold without boundary, or a manifold embedded in another manifold with boundary. Think of a closed interval embedded in an ordinary half plane. Two cases arise. The case where the interval lies inside the interior of the half plane is essentially distinct from the case where the interval has one end point touching the hyperplane forming the boundary of the half plane. (For instance, given two embeddings of the first type, there exists an automorphism of the half plane carrying one into the other, but there cannot exist an automorphism of the half plane carrying an embedding of the first type into one of the second type.)

We leave it to the reader to go systematically through the notions of tangent space, immersion, embedding (and later, tangent bundle, vector field, etc.) for arbitrary manifolds (with boundary). For instance, Proposition 1 shows at once how to get the tangent space functorially.

CHAPTER III

Vector Bundles

The collection of tangent spaces can be glued together to give a manifold with a natural projection, thus giving rise to the tangent bundle. The general glueing procedure can be used to construct more general objects known as vector bundles, which give powerful invariants of a given manifold. (For an interesting theorem see Mazur [14].) In this chapter, we develop purely formally certain functorial constructions having to do with vector bundles. In the chapters on differential forms and Riemannian metrics, we shall discuss in greater detail the constructions associated with multilinear alternating forms, and symmetric positive definite forms.

Partitions of unity are an essential tool when considering vector bundles. They can be used to combine together a random collection of morphisms into vector bundles, and we shall give a few examples showing how this can be done (concerning exact sequences of bundles).

§1. Definition, pull-backs

Let X be a manifold (of class C^p with $p \geqq 0$) and let $\pi: E \to X$ be a morphism. Assume:

VB 1. *For each $x \in X$, the fiber $\pi^{-1}(x) = E_x$ has been given the structure of a Banach space.*

Let $\{U_i\}$ be an open covering of X, and for each i, suppose that we are given a Banach space $\mathbf{E}_i$ and a mapping

$$\tau_i: \pi^{-1}(U_i) \to U_i \times \mathbf{E}_i$$

satisfying the following conditions:

VB 2. *The map τ_i is an isomorphism commuting with the projection on U_i, that is, such that the following diagram is commutative:*

$$\pi^{-1}(U_i) \xrightarrow{\ \tau_i\ } U_i \times \mathbf{E}_i$$
$$\searrow \qquad \swarrow$$
$$U_i$$

41

and for each $x \in U_i$ the induced map on the fiber (written $\tau_i(x)$ or τ_{ix})

$$\tau_{ix} \colon \pi^{-1}(x) \to \mathbf{E}_i$$

is a toplinear isomorphism.

VB 3. *If U_i and U_j are two members of the covering, then the map of $U_i \cap U_j$ into $L(\mathbf{E}_i, \mathbf{E}_j)$ given by*

$$x \mapsto (\tau_j \tau_i^{-1})_x$$

is a morphism.

Then we shall say that $\{(U_i, \tau_i)\}$ is a **trivialising covering** for π (or for E by abuse of language), and that $\{\tau_i\}$ are its **trivialising** maps. If $x \in U_i$, we say that τ_i (or U_i) trivialises at x. Two trivialising coverings for π are said to be **VB-equivalent** if taken together they also satisfy conditions **VB 2**, **VB 3**. An equivalence class of such trivialising coverings is said to determine a structure of **vector bundle** on π (or on E by abuse of language). We say that E is the **total space** of the bundle, and that X is its **base space**. If we wish to be very functorial, we shall write E_π and X_π for these spaces respectively. The fiber $\pi^{-1}(x)$ is also denoted by E_x or π_x.

If all the spaces $\mathbf{E}_i$ are toplinearly isomorphic, then one could assume that they are all equal, say to the space $\mathbf{E}$. In that case we say that π (or E by abuse of language) is a **vector bundle with fiber E**. If X is connected, this is necessarily the case, because the set of points $x \in X$ for which there is a trivialising map

$$\tau_{ix} \colon \pi^{-1}(x) \to \mathbf{E}$$

with a given space $\mathbf{E}$ is both open and closed.

In the finite dimensional case, condition **VB 3** is implied by **VB 2**.

Proposition 1. *Let $\mathbf{E}, \mathbf{F}$ be finite dimensional vector spaces. Let U be open in some Banach space. Let*

$$f \colon U \times \mathbf{E} \to \mathbf{F}$$

be a morphism such that for each $x \in U$, the map

$$f_x \colon \mathbf{E} \to \mathbf{F}$$

given by $f_x(v) = f(x, v)$ is a linear map. Then the map of U into $L(\mathbf{E}, \mathbf{F})$ given by $x \mapsto f_x$ is a morphism.

Proof. We can write $\mathbf{F} = \mathbf{R}_1 \times \cdots \times \mathbf{R}_n$ (n copies of $\mathbf{R}$). Using the fact that $L(\mathbf{E}, \mathbf{F}) = L(\mathbf{E}, \mathbf{R}_1) \times \cdots \times L(\mathbf{E}, \mathbf{R}_n)$, it will suffice to prove our assertion when $\mathbf{F} = \mathbf{R}$. Similarly, we can assume that $\mathbf{E} = \mathbf{R}$ also. But in that case, the function $f(x, v)$ can be written $g(x)v$ for some map $g \colon U \to \mathbf{R}$. Since f is a morphism, it follows that as a function of each argument x, v it is also a morphism. Putting $v = 1$ shows that g is a morphism and concludes the proof.

As with manifolds, we need less than our three properties to recover vector bundles.

Proposition 2. *Let X be a manifold, and $\pi \colon E \to X$ a mapping from some set E into X. Let $\{U_i\}$ be an open covering of X, and for each i suppose that we are given a Banach space $\mathbf{E}_i$ and a bijection (commuting with the projection on U_i),*

$$\tau_i \colon \pi^{-1}(U_i) \to U_i \times \mathbf{E}_i,$$

such that for each pair i, j and $x \in U_i \cap U_j$, the map $(\tau_j \tau_i^{-1})_x$ is a toplinear isomorphism, and condition **VB 3** *is satisfied. Then there exists a unique structure of manifold on E such that π is a morphism, such that τ_i is an isomorphism making π into a vector bundle, and $\{(U_i, \tau_i)\}$ into a trivialising covering.*

Proof. By Proposition 16 of Chapter I, §3 and our condition **VB 3** we conclude that the map

$$\tau_j \tau_i^{-1} \colon (U_i \cap U_j) \times \mathbf{E}_i \to (U_i \cap U_j) \times \mathbf{E}_j$$

is a morphism, and in fact an isomorphism since it has an inverse. From the definition of atlases, we conclude that E has a unique manifold structure such that the τ_i are isomorphisms. Since π is obtained locally as a composite of morphisms (namely τ_i and the projections of $U_i \times \mathbf{E}_i$ on the first factor), it becomes a morphism. On each fiber $\pi^{-1}(x)$, we can transport the topological vector space structure of any $\mathbf{E}_i$ such that x lies in U_i, by means of τ_{ix}. The result is independent of the choice of U_i since $(\tau_j \tau_i^{-1})_x$ is a toplinear isomorphism. Our proposition is proved.

We now make the set of vector bundles into a category.
Let $\pi \colon E \to X$ and $\pi' \colon E' \to X'$ be two vector bundles.
A **VB-morphism** $\pi \to \pi'$ consists of a pair of morphisms

$$f_0 \colon X \to X' \quad \text{and} \quad f \colon E \to E'$$

satisfying the following conditions.

VB Mor 1. *The diagram*

$$
\begin{array}{ccc}
E & \xrightarrow{\ f\ } & E' \\
\pi \downarrow & & \downarrow \pi' \\
X & \xrightarrow{\ f_0\ } & X'
\end{array}
$$

is commutative, and the induced map for each $x \in X$

$$f_x : E_x \to E'_{f(x)}$$

is a continuous linear map.

VB Mor 2. *For each $x_0 \in X$ there exist trivialising maps*

$$\tau : \pi^{-1}(U) \to U \times \mathbf{E}$$

and

$$\tau' : \pi'^{-1}(U') \to U' \times \mathbf{E}'$$

at x_0 and $f(x_0)$ respectively, such that $f_0(U)$ is contained in U', and such that the map of U into $L(\mathbf{E}, \mathbf{E}')$ given by

$$x \mapsto \tau'_{f_0(x)} \circ f_x \circ \tau^{-1}$$

is a morphism.

As a matter of notation, we shall also use f to denote the VB-morphism, and thus write $f : \pi \to \pi'$. In most applications, f_0 is the identity. By Proposition 1, we observe that **VB Mor 2** is redundant in the finite dimensional case.

The next proposition is the analogue of Proposition 2 for VB-morphisms.

Proposition 3. *Let π, π' be two vector bundles over manifolds X, X' respectively. Let $f_0 : X \to X'$ be a morphism, and suppose that we are given for each $x \in X$ a continuous linear map*

$$f_x : \pi_x \to \pi'_{f_0(x)}$$

*such that, for each x_0, condition **VB Mor 2** is satisfied. Then the map f from π to π' defined by f_x on each fiber is a VB-morphism.*

Proof. One must first check that f is a morphism. This can be done under the assumption that π, π' are trivial, say equal to $U \times \mathbf{E}$ and $U' \times \mathbf{E}'$ (following the notation of **VB Mor 2**), with trivialising maps equal to the identity. Our map f is then given by

$$(x, v) \mapsto (f_0 x, f_x v).$$

Using Proposition 16 of Chapter I, §3 we conclude that f is a morphism, and hence that (f_0, f) is a VB-morphism.

It is clear how to compose two VB-morphisms set theoretically. In fact, the composite of two VB-morphisms is a VB-morphism. There is no problem verifying condition **VB Mor 1**, and for **VB Mor 2**, we look at the situation locally. We encounter a commutative diagram of the following type:

$$
\begin{array}{ccc}
\pi^{-1}(U) \xrightarrow{\;f\;} \pi'^{-1}(U') \xrightarrow{\;g\;} \pi''^{-1}(U'') \\
\Big\downarrow{\tau} \qquad\qquad \Big\downarrow{\tau'} \qquad\qquad \Big\downarrow{\tau''} \\
U \times \mathbf{E} \longrightarrow U' \times \mathbf{E}' \longrightarrow U'' \times \mathbf{E}''
\end{array}
$$

and use Proposition 16 of Chapter I, §3 to show that $g \circ f$ is a VB-morphism.

We therefore have a category, denoted by VB or VB^p, if we need to specify explicitly the order of differentiability.

The vector bundles over X form a subcategory $\mathrm{VB}(X) = \mathrm{VB}^p(X)$ (taking those VB-morphisms for which the map f_0 is the identity). If $\mathfrak{A}$ is a category of Banach spaces (for instance finite dimensional spaces), then we denote by $\mathrm{VB}(X, \mathfrak{A})$ those vector bundles over X whose fibers lie in $\mathfrak{A}$.

A morphism from one vector bundle into another can be given locally. More precisely, suppose that U is an open subset of X and $\pi : E \to X$ a vector bundle over X. Let $E_U = \pi^{-1}(U)$ and

$$
\pi_U = \pi \mid E_U
$$

be the restriction of π to E_U. Then π_U is a vector bundle over U. Let $\{U_i\}$ be an open covering of the manifold X and let π, π' be two vector bundles over X. Suppose, given a VB-morphism

$$
f_i : \pi_{U_i} \to \pi'_{U_i}
$$

for each i, such that f_i and f_j agree over $U_i \cap U_j$ for each pair of indices i, j. Then there exists a unique VB-morphism $f : \pi \to \pi'$ which agrees with f_i on each U_i. The proof is trivial, but the remark will be used frequently in the sequel.

Using the discussion at the end of Chapter II, §2 and Proposition 6 of that chapter, we get immediately:

Proposition 4. *Let* $\pi : E \to Y$ *be a vector bundle, and* $f : X \to Y$ *a morphism. Then*

$$
f^*(\pi) : f^*(E) \to X
$$

is a vector bundle, and the pair $(f, \pi^*(f))$ *is a* VB-*morphism.*

$$f^*(E) \xrightarrow{\pi^*(f)} E$$
$$f^*(\pi) \downarrow \qquad\qquad \downarrow \pi$$
$$X \xrightarrow{\quad f \quad} Y$$

In Proposition 4, we could take f to be the inclusion of a submanifold. In that case, the pull-back is merely the restriction. As with open sets, we can then use the usual notation:

$$E_X = \pi^{-1}(X) \qquad \text{and} \qquad \pi_X = \pi \mid E_X.$$

Thus $\pi_X = f^*(\pi)$ in that case.

If X happens to be a point y of Y, then we have the constant map

$$\pi_y \colon E_y \to y$$

which will sometimes be identified with E_y.

If we identify each fiber $(f^*E)_x$ with $E_{f(x)}$ itself (a harmless identification since an element of the fiber at x is simply a pair (x, e) with e in $E_{f(x)}$), then we can describe the pull-back f^* of a vector bundle $\pi\colon E \to Y$ as follows. It is a vector bundle $f^*\pi\colon f^*E \to X$ satisfying the following properties.

PB 1. *For each* $x \in X$, *we have* $(f^*E)_x = E_{f(x)}$.

PB 2. *We have a commutative diagram*

$$f^*E \longrightarrow E$$
$$f^*\pi \downarrow \qquad\qquad \downarrow \pi$$
$$X \xrightarrow{\quad f \quad} Y$$

the top horizontal map being the identity on each fiber.

PB 3. *If* E *is trivial, equal to* $Y \times \mathbf{E}$, *then* $f^*E = X \times \mathbf{E}$ *and* $f^*\pi$ *is the projection.*

PB 4. *If* V *is an open subset of* Y *and* $U = f^{-1}(V)$, *then*

$$f^*(E_V) = (f^*E)_U,$$

and we have a commutative diagram.

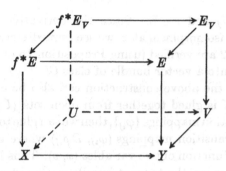

§2. *The tangent bundle*

Let X be a manifold of class C^p with $p \geqq 1$. We shall define a functor T from the category of such manifolds into the category of vector bundles of class C^{p-1}.

For each manifold X we let $T(X)$ be the disjoint union of the tangent spaces $T_x(X)$. We have a natural projection

$$\pi \colon T(X) \to X$$

mapping $T_x(X)$ on x. We must make this into a vector bundle. If (U, φ) is a chart of X such that φU is open in the Banach space $\mathbf{E}$, then from the definition of the tangent vectors as equivalence classes of triples (U, φ, v) we get immediately a bijection

$$\tau_U \colon \pi^{-1}(U) = T(U) \to U \times \mathbf{E}$$

which commutes with the projection on U, that is such that

$$\pi^{-1}(U) \xrightarrow{\ \tau_U\ } U \times \mathbf{E}$$
$$\searrow \quad \swarrow$$
$$U$$

is commutative. Furthermore, if (U_i, φ_i) and (U_j, φ_j) are two charts, and if we denote by φ_{ji} the map $\varphi_j\varphi_i^{-1}$ (defined on $\varphi_i(U_i \cap U_j)$), then we obtain a transition mapping

$$\tau_{ji} = (\tau_j\tau_i^{-1})\colon \varphi_i(U_i \cap U_j) \times \mathbf{E} \to \varphi_j(U_i \cap U_j) \times \mathbf{E}$$

by the formula

$$\tau_{ji}(x, v) = (\varphi_{ji}x, D\varphi_{ji}(x) \cdot v)$$

for $x \in U_i \cap U_j$ and $v \in \mathbf{E}$. Since the derivative $D\varphi_{ji} = \varphi'_{ji}$ is of class C^{p-1} and is an isomorphism at x, we see immediately that all the conditions of Proposition 2 are verified (using Proposition 16 of Chapter I, §3), thereby making $T(X)$ into a vector bundle of class C^{p-1}.

We see that the above construction can also be expressed as follows. If the manifold X is glued together from open sets $\{U_i\}$ in Banach spaces by means of transition mappings $\{\varphi_{ij}\}$, then we can glue together products $U_i \times \mathbf{E}$ by means of transition mappings $(\varphi_{ij}, D\varphi_{ij})$ where the derivative $D\varphi_{ij}$ can be viewed as a function of two variables (x, v). Thus locally, for open subsets U of Banach spaces, the tangent bundle can be identified with the product $U \times \mathbf{E}$. The reader will note that our definition coincides with the oldest definition employed by geometers, our tangent vectors being vectors which transform according to a certain rule (namely the derivative).

If $f: X \to X'$ is a C^p-morphism, we can define

$$Tf: T(X) \to T(X')$$

to be simply $T_x f$ on each fiber $T_x(X)$. In order to verify that Tf is a VB-morphism (of class C^{p-1}), it suffices to look at the situation locally, i.e. we may assume that X and X' are open in vector spaces $\mathbf{E}, \mathbf{E}'$, and that $T_x f = f'(x)$ is simply the derivative. Then the map Tf is given by

$$Tf(x, v) = (f(x), f'(x)v)$$

for $x \in X$ and $v \in \mathbf{E}$. Since f' is of class C^{p-1} by definition, we can apply Proposition 16 loc. cit. to conclude that Tf is also of class C^{p-1}. The functoriality property is trivially satisfied, and we have therefore defined the functor T as promised.

It will sometimes be notationally convenient to write f_* instead of Tf for the induced map, which is also called the **tangent** map. The bundle $T(X)$ is called the **tangent bundle** of X.

§3. Exact sequences of bundles

Let X be a manifold. Let $\pi': E' \to X$ and $\pi: E \to X$ be two vector bundles over X. Let $f: \pi' \to \pi$ be a VB-morphism. We shall say that the sequence

$$0 \to \pi' \xrightarrow{\ f\ } \pi$$

is **exact** if there exists a covering of X by open sets and for each open set U in this covering there exist trivialisations

$$\tau': E'_U \to U \times \mathbf{E}' \qquad \text{and} \qquad \tau: E_U \to U \times \mathbf{E}$$

such that $\mathbf{E}$ can be written as a product $\mathbf{E} = \mathbf{E}' \times \mathbf{F}$, making the following diagram commutative:

$$
\begin{array}{ccc}
E'_U & \xrightarrow{\;f\;} & E_U \\
\tau' \downarrow & & \downarrow \tau \\
U \times \mathbf{E}' & \longrightarrow & U \times \mathbf{E}' \times \mathbf{F}
\end{array}
$$

(The bottom map is the natural one: Identity on U and the injection of $\mathbf{E}'$ on $\mathbf{E}' \times 0$.)

Let $\pi_1 \colon E_1 \to X$ be another vector bundle, and let $g \colon \pi_1 \to \pi$ be a VB-morphism such that $g(E_1)$ is contained in $f(E')$. Since f establishes a bijection between E' and its image $f(E')$ in E, it follows that there exists a unique map $g_1 \colon E_1 \to E'$ such that $g = f \circ g_1$. We contend that g_1 is a VB-morphism. Indeed, to prove this we can work locally, and in view of the definition, over an open set U as above, we can write

$$
g_1 = \tau'^{-1} \circ \mathrm{pr} \circ \tau \circ g
$$

where pr is the projection of $U \times \mathbf{E}' \times \mathbf{F}$ on $U \times \mathbf{E}'$. All the maps on the right-hand side of our equality are VB-morphisms; this proves our contention.

Let $\pi \colon E \to X$ be a vector bundle. A subset S of E will be called a **subbundle** if there exists an exact sequence $0 \to \pi' \to \pi$, also written

$$
0 \to E' \xrightarrow{\;f\;} E,
$$

such that $f(E') = S$. This gives S the structure of a vector bundle, and the previous remarks show that it is unique. In fact, given another exact sequence

$$
0 \to E_1 \xrightarrow{\;g\;} E
$$

such that $g(E_1) = S$, the natural map $f^{-1}g$ from E_1 to E' is a VB-isomorphism.

Let us denote by E/E' the union of all factor spaces E_x/E'_x. If we are dealing with an exact sequence as above, then we can give E/E' the structure of a vector bundle. We proceed as follows. Let $\{U_i\}$ be our covering, with trivialising maps τ'_i and τ_i. We can define for each i a bijection

$$
\tau''_i \colon E_{U_i}/E'_{U_i} \to U_i \times \mathbf{F}
$$

obtained in a natural way from the above commutative diagram. (Without loss of generality, we can assume that the vector spaces $\mathbf{E}'$, $\mathbf{F}$ are constant for all i.) We have to prove that these bijections satisfy the conditions of Proposition 2.

Without loss of generality, we may assume that f is an inclusion (of the total space E' into E). For each pair i, j and $x \in U_i \cap U_j$, the toplinear automorphism $(\tau_j \tau_i^{-1})_x$ is represented by a matrix

$$\begin{pmatrix} h_{11}(x) & h_{12}(x) \\ h_{21}(x) & h_{22}(x) \end{pmatrix}$$

operating on the right on a vector $(v, w) \in \mathbf{E}' \times \mathbf{F}$. The map $(\tau_j'' \tau_i''^{-1})_x$ on $\mathbf{F}$ is induced by this matrix. Since $\mathbf{E}' = \mathbf{E}' \times 0$ has to be carried into itself by the matrix, we have $h_{12}(x) = 0$. Furthermore, since $(\tau_j \tau_i^{-1})_x$ has an inverse, equal to $(\tau_i \tau_j^{-1})_x$, it follows that $h_{22}(x)$ is a toplinear automorphism of $\mathbf{F}$, and represents $(\tau_j'' \tau_i''^{-1})_x$. Therefore condition **VB 3** is satisfied, and E/E' is a vector bundle.

The canonical map

$$E_U \to E_U/E_U'$$

is a morphism since it can be expressed in terms of τ, the projection, and τ''^{-1}. Consequently, we obtain a VB-morphism

$$g: \pi \to \pi''$$

in the canonical way (on the total spaces, it is the quotient mapping of E on E/E'). We shall call π'' the **factor bundle**.

Our map g satisfies the usual universal mapping property of a cokernel. Indeed, suppose that

$$\psi: E \to G$$

is a VB-morphism such that $\psi \circ f = 0$ (i.e. $\psi_x \circ f_x = 0$ on each fiber E_x'). We can then define set theoretically a canonical map

$$\psi_*: E/E' \to G,$$

and we must prove that it is a VB-morphism. This can be done locally. Using the above notation, we may assume that $E = U \times \mathbf{E}' \times \mathbf{F}$ and that g is the projection. In that case, ψ_* is simply the canonical injection of $U \times \mathbf{F}$ in $U \times \mathbf{E}' \times \mathbf{F}$ followed by ψ, and is therefore a VB-morphism.

We shall therefore call g the **cokernel** of f.

Dually, let $g: \pi \to \pi''$ be a given VB-morphism. We shall say that the sequence

$$\pi \xrightarrow{g} \pi'' \to 0$$

is **exact** if g is surjective, and if there exists a covering of X by open sets, and for each open set U in this covering there exist spaces $\mathbf{E'}, \mathbf{F}$ and trivialisations

$$\tau : E_U \to U \times \mathbf{E'} \times \mathbf{F} \qquad \text{and} \qquad \tau'' : E_U'' \to \mathbf{F}$$

making the following diagram commutative.

$$
\begin{array}{ccc}
E_U & \xrightarrow{\;\;g\;\;} & E_U'' \\
\tau \downarrow & & \downarrow \tau'' \\
U \times \mathbf{E'} \times \mathbf{F} & \longrightarrow & U \times \mathbf{F}
\end{array}
$$

(The bottom map is the natural one: Identity on U and the projection of $\mathbf{E'} \times \mathbf{F}$ on $\mathbf{F}$.)

In the same way as before, one sees that the "kernel" of g, that is, the union of the kernels E_x' of each g_x, can be given a structure of vector bundle. This union E' will be called the **kernel** of g, and satisfies the usual universal mapping property.

Proposition 5. *Let X be a manifold and let*

$$f : \pi' \to \pi$$

be a VB-morphism of vector bundles over X. Assume that, for each $x \in X$, the continuous linear map

$$f_x : E_x' \to E_x$$

is injective and splits. Then the sequence

$$0 \to \pi' \xrightarrow{\;\;f\;\;} \pi$$

is exact.

Proof. We can assume that X is connected and that the fibers of E' and E are constant, say equal to the Banach spaces $\mathbf{E'}$ and $\mathbf{E}$. Let $a \in X$. Corresponding to the splitting of f_a we know that we have a product decomposition $\mathbf{E} = \mathbf{E'} \times \mathbf{F}$ and that there exists an open set U of X containing a, together with trivialising maps

$$\tau : \pi^{-1}(U) \to U \times \mathbf{E} \qquad \text{and} \qquad \tau' : \pi'^{-1}(U) \to U \times \mathbf{E'}$$

such that the composite map

$$E' \xrightarrow{\tau_a'^{-1}} E_a' \xrightarrow{f_a} E_a \xrightarrow{\tau_a} E' \times F$$

maps E' on $E' \times 0$.

For any point x in U, we have a map

$$(\tau f \tau'^{-1})_x \colon E' \to E' \times F,$$

which can be represented by a pair of continuous linear maps $\big(h_{11}(x), h_{21}(x)\big)$. We define

$$h(x) \colon E' \times F \to E' \times F$$

by the matrix

$$\begin{pmatrix} h_{11}(x) & 0 \\ h_{21}(x) & id \end{pmatrix},$$

operating on the right on a vector $(v, w) \in E' \times F$. Then $h(x)$ restricted to $E' \times 0$ has the same action as $(\tau f \tau'^{-1})_x$.

The map $x \to h(x)$ is a morphism of U into $L(E, E)$ and since it is continuous, it follows that for U small enough around our fixed point a, it maps U into the group of toplinear automorphisms of E. This proves our proposition.

Dually to Proposition 5, we have:

Proposition 6. *Let X be a manifold and let*

$$g \colon \pi \to \pi''$$

be a VB-morphism of vector bundles over X. Assume that for each $x \in X$, the continuous linear map

$$g_x \colon E_x \to E_x''$$

is surjective and has a kernel that splits. Then the sequence

$$\pi \xrightarrow{g} \pi'' \to 0$$

is exact.

Proof. It is dual to the preceding one and we leave it to the reader.

In general, a sequence of VB-morphisms

$$0 \to \pi' \xrightarrow{f} \pi \xrightarrow{g} \pi'' \to 0$$

is said to be **exact** if both ends are exact, and if the image of f is equal to the kernel of g.

There is an important example of exact sequence. Let $f: X \to Y$ be an immersion. By the universal mapping property of pull-backs, we have a canonical VB-morphism

$$T^*f: T(X) \to f^*T(Y)$$

of $T(X)$ into the pull-back over X of the tangent bundle of Y. Furthermore, from the manner in which the pull-back is obtained locally by taking products, and the definition of an immersion, one sees that the sequence

$$0 \to T(X) \xrightarrow{T^*f} f^*T(Y)$$

is exact. The factor bundle

$$f^*T(Y)/\mathrm{Im}\ (T^*f)$$

is called the **normal** bundle of f. It is denoted by $N(f)$, and its total space by $N_f(X)$ if we wish to distinguish between the two. We sometimes identify $T(X)$ with its image under T^*f and write

$$N(f) = f^*T(Y)/T(X).$$

Dually, let $f: X \to Y$ be a submersion. Then we have an exact sequence

$$T(X) \xrightarrow{T^*f} f^*T(Y) \to 0$$

whose kernel could be called the **subbundle** of f, or the **bundle along the fiber**.

There is an interesting case where we can describe the kernel more precisely. Let

$$\pi: E \to X$$

be a vector bundle. Then we can form the pull-back of E over itself, that is, π^*E, and we contend that we have an exact sequence

$$0 \to \pi^*E \to T(E) \to \pi^*T(X) \to 0.$$

To define the map on the left, we look at the subbundle of π more closely. For each $x \in X$ we have an inclusion

$$E_x \to E,$$

whence a natural injection

$$T(E_x) \to T(E).$$

The local product structure of a bundle shows that the union of the $T(E_x)$ as x ranges over X gives the subbundle set theoretically. On the other hand, the total space of π^*E consists of pairs of vectors (v, w) lying over the same base point x, that is, the fiber at x of $\pi^* E$ is simply $E_x \times E_x$. Since $T(E_x)$ has a natural identification with $E_x \times E_x$, we get for each x a bijection

$$(\pi^* E)_x \to T(E_x)$$

which defines our map from $\pi^* E$ to $T(E)$. Considering the map locally in terms of the local product structure shows at once that it gives a VB-isomorphism between $\pi^* E$ and the subbundle of π, as desired.

§4. Operations on vector bundles

We consider subcategories of Banach spaces $\mathfrak{A}$, $\mathfrak{B}$, $\mathfrak{C}$ and let

$$\lambda: \mathfrak{A} \times \mathfrak{B} \to \mathfrak{C}$$

be a functor in two variables, which is, say, contravariant in the first and covariant in the second. (Everything we shall do extends in the obvious manner to functors of several variables, letting $\mathfrak{A}$, $\mathfrak{B}$ stand for n-tuples.)

If $f: \mathbf{E}' \to \mathbf{E}$ and $g: \mathbf{F} \to \mathbf{F}'$ are two continuous linear maps, with f a morphism of $\mathfrak{A}$ and g a morphism of $\mathfrak{B}$, then by definition, we have a map

$$L(\mathbf{E}', \mathbf{E}) \times L(\mathbf{F}, \mathbf{F}') \to L(\lambda(\mathbf{E}, \mathbf{F}), \lambda(\mathbf{E}', \mathbf{F}')),$$

assigning $\lambda(f, g)$ to (f, g).

We shall say that λ is of **class** C^p if the following condition is satisfied. Given a manifold U, and two morphisms

$$\varphi: U \to L(\mathbf{E}', \mathbf{E}) \qquad \text{and} \qquad \psi: U \to L(\mathbf{F}, \mathbf{F}'),$$

then the composite

$$U \to L(\mathbf{E}', \mathbf{E}) \times L(\mathbf{F}, \mathbf{F}') \to L(\lambda(\mathbf{E}, \mathbf{F}), \lambda(\mathbf{E}', \mathbf{F}'))$$

is also a morphism. (One could also say that λ is **differentiable**.)

Theorem 1. *Let λ be a functor as above, of class C^p, $p \geqq 0$. Then for each manifold X, there exists a functor λ_X, on vector bundles (of class C^p)*

$$\lambda_X: \mathrm{VB}(X, \mathfrak{A}) \times \mathrm{VB}(X, \mathfrak{B}) \to \mathrm{VB}(X, \mathfrak{C})$$

satisfying the following properties. For any bundles α, β in $\mathrm{VB}(X, \mathfrak{A})$ and $\mathrm{VB}(X, \mathfrak{B})$ respectively, and VB-morphisms

$$f: \alpha' \to \alpha \qquad \text{and} \qquad g: \beta \to \beta'$$

in the respective categories, and for each $x \in X$, we have:

OP 1. $\lambda_X(\alpha, \beta)_x = \lambda(\alpha_x, \beta_x)$.

OP 2. $\lambda_X(f, g)_x = \lambda(f_x, g_x)$.

OP 3. *If α is the trivial bundle $X \times \mathbf{E}$ and β the trivial bundle $X \times \mathbf{F}$, then $\lambda_X(\alpha, \beta)$ is the trivial bundle $X \times \lambda(\mathbf{E}, \mathbf{F})$.*

OP 4. *If $h: Y \to X$ is a C^p-morphism, then*

$$\lambda_Y^*(h^*\alpha, h^*\beta) = h^*\lambda_X(\alpha, \beta).$$

Proof. We may assume that X is connected, so that all the fibers are toplinearly isomorphic to a fixed space. For each open subset U of X we let the total space $\lambda_U(E_\alpha, E_\beta)$ of $\lambda_U(\alpha, \beta)$ be the union of the sets

$$\{x\} \times \lambda(\alpha_x, \beta_x)$$

(identified harmlessly throughout with $\lambda(\alpha_x, \beta_x)$), as x ranges over U. We can find a covering $\{U_i\}$ of X with trivialising maps $\{\tau_i\}$ for α, and $\{\sigma_i\}$ for β,

$$\tau_i: \alpha^{-1}(U_i) \to U_i \times \mathbf{E}$$

$$\sigma_i: \beta^{-1}(U_i) \to U_i \times \mathbf{F}.$$

We have a bijection

$$\lambda(\tau_i^{-1}, \sigma_i): \lambda_{U_i}(E_\alpha, E_\beta) \to U_i \times \lambda(\mathbf{E}, \mathbf{F})$$

obtained by taking on each fiber the map

$$\lambda(\tau_{ix}^{-1}, \sigma_{ix}): \lambda(\alpha_x, \beta_x) \to \lambda(\mathbf{E}, \mathbf{F}).$$

We must verify that **VB 3** is satisfied. This means looking at the map

$$x \to \lambda(\tau_{jx}^{-1}, \sigma_{jx}) \circ \lambda(\tau_{ix}^{-1}, \sigma_{ix})^{-1}.$$

The expression on the right is equal to

$$\lambda(\tau_{ix}\tau_{jx}^{-1}, \sigma_{jx}\sigma_{ix}^{-1}).$$

Since λ is a functor of class C^p, we see that we get a map

$$U_i \cap U_j \to L\big(\lambda(\mathbf{E}, \mathbf{F}), \lambda(\mathbf{E}, \mathbf{F})\big)$$

which is a C^p-morphism. Furthermore, since λ is a functor, the transition mappings are in fact toplinear isomorphisms, and **VB 2**, **VB 3** are proved.

The proof of the analogous statement for $\lambda_X(f, g)$, to the effect that it is a VB-morphism, proceeds in an analogous way, again using the hypothesis that λ is of class C^p. Condition **OP 3** is obviously satisfied, and **OP 4** follows by localizing. This proves our theorem.

The next theorem gives us the uniqueness of the operation λ_X.

Theorem 2. *If μ is another functor of class C^p with the same variance as λ, and if we have a natural transformation of functors $t: \lambda \to \mu$, then for each X, the mapping*

$$t_X: \lambda_X \to \mu_X,$$

defined on each fiber by the map

$$t(\alpha_x, \beta_x): \lambda(\alpha_x, \beta_x) \to \mu(\alpha_x, \beta_x),$$

is a natural transformation of functors (in the VB-category).

Proof. For simplicity of notation, assume that λ and μ are both functors of one variable, and both covariant. For each open set $U = U_i$ of a trivialising covering for β, we have a commutative diagram.

$$
\begin{array}{ccc}
U \times \lambda(\mathbf{E}) & \xrightarrow{\ \mathrm{id} \times t(\mathbf{E})\ } & U \times \mu(\mathbf{E}) \\[4pt]
{\scriptstyle \lambda_U(\sigma)} \uparrow & & \uparrow {\scriptstyle \mu_U(\sigma)} \\[4pt]
\lambda_U(\beta) & \xrightarrow[\ t_U\ ]{} & \mu_U(\beta)
\end{array}
$$

The vertical maps are trivialising VB-isomorphisms, and the top horizontal map is a VB-morphism. Hence t_U is a VB-morphism, and our assertion is proved.

In particular, for $\lambda = \mu$ and $t = id$ we get the uniqueness of our functor λ_X.

(In the proof of Theorem 2, we do not use again explicitly the hypotheses that λ, μ are differentiable.)

In practice, we omit the subscript X on λ, and write λ for the functor on vector bundles.

As an example of operation, we have the **direct sum** (also called the **Whitney sum**) of two bundles α, β over X. It is denoted by $\alpha \oplus \beta$, and the fiber at a point x is

$$(\alpha \oplus \beta)_x = \alpha_x \oplus \beta_x.$$

Of course, the finite direct sum of vector spaces can be identified with their finite direct products, but we write the above operation as a direct sum in order not to confuse it with the following direct product.

Let $\alpha: E_\alpha \to X$ and $\beta: E_\beta \to Y$ be two vector bundles in $VB(X)$ and $VB(Y)$ respectively. Then the map

$$\alpha \times \beta: E_\alpha \times E_\beta \to X \times Y$$

is a vector bundle, and it is this operation which we call the **direct product** of α and β.

Let X be a manifold, and λ a functor of class C^p with $p \geqq 1$. The **tensor bundle** of type λ over X is defined to be $\lambda_X(T(X))$, also denoted by $\lambda T(X)$ or $T_\lambda(X)$. The sections of this bundle are called **tensor fields** of type λ, and the set of such sections is denoted by $\Gamma_\lambda(X)$. Suppose that we have a trivialisation of $T(X)$, say

$$T(X) = X \times \mathbf{E}.$$

Then $T_\lambda(X) = X \times \lambda(\mathbf{E})$. A section of $T_\lambda(X)$ in this representation is completely described by the projection on the second factor, which is a morphism

$$f: X \to \lambda(\mathbf{E}).$$

We shall call it the **local representation** of the tensor field (in the given trivialisation). If ξ is the tensor field having f as its local representation, then

$$\xi(x) = \big(x, f(x)\big).$$

Let $f: X \to Y$ be a morphism of class C^p $(p \geqq 1)$. Let ω be a tensor field of type L^r over Y, which could also be called a multilinear tensor field. For each $y \in Y$, $\omega(y)$ (also written ω_y) is a continuous multilinear function on $T_y(Y)$:

$$\omega_y: T_y \times \cdots \times T_y \to \mathbf{R}.$$

For each $x \in X$, we can define a continuous multilinear map

$$f_x^*(\omega): T_x \times \cdots \times T_x \to \mathbf{R}$$

by the composition of maps $(T_x f)^r$ and $\omega_{f(x)}$:

$$T_x \times \cdots \times T_x \to T_{f(x)} \times \cdots \times T_{f(x)} \to \mathbf{R}.$$

We contend that the map $x \mapsto f_x^*(\omega)$ is a tensor field over X, of the same type as ω. To prove this, we may work with local representation. Thus we can assume that we work with a morphism

$$f: U \to V$$

of one open set in a Banach space into another, and that

$$\omega \colon V \to L'(\mathbf{F})$$

is a morphism, V being open in $\mathbf{F}$. If U is open in $\mathbf{E}$, then $f^*(\omega)$ (now denoting a local representation) becomes a mapping of U into $L'(\mathbf{E})$, given by the formula

$$f_x^*(\omega) = L'(f'(x)) \cdot \omega(f(x)).$$

Since $L' \colon L(\mathbf{E}, \mathbf{F}) \to L(L'(\mathbf{F}), L'(\mathbf{E}))$ is of class C^∞, it follows that $f^*(\omega)$ is a morphism of the same class as ω. This proves what we want.

Of course, the same argument is valid for the other functors L_s^r and L_a^r (symmetric and alternating continuous multilinear maps). Special cases will be considered in later chapters. If λ denotes any one of our three functors, then we see that we have obtained a mapping (which is in fact linear)

$$f^* \colon \Gamma_\lambda(Y) \to \Gamma_\lambda(X)$$

which is clearly functorial in f. We use the notation f^* instead of the more correct (but clumsy) notation f_λ or $\Gamma_\lambda(f)$. No confusion will arise from this.

§5. *Splitting of vector bundles*

The next proposition expresses the fact that the VB-morphisms of one bundle into another (over a fixed morphism) form a module over the ring of functions.

Proposition 7. *Let X, Y be manifolds and $f_0 \colon X \to Y$ a morphism. Let α, β be vector bundles over X, Y respectively, and let $f, g \colon \alpha \to \beta$ be two VB-morphisms over f_0. Then the map $f + g$ defined by the formula*

$$(f + g)_x = f_x + g_x$$

is also a VB-morphism. Furthermore, if $\psi \colon Y \to \mathbf{R}$ is a function on Y, then the map ψf defined by

$$(\psi f)_x = \psi(f_0(x))f_x$$

is also a VB-morphism.

Proof. Both assertions are immediate consequences of Proposition 16, Chapter I, §3.

We shall consider mostly the situation where $X = Y$ and f_0 is the identity, and will use it, together with partitions of unity, to glue VB-morphisms together.

Let α, β be vector bundles over X and let $\{(U_i, \psi_i)\}$ be a partition of unity on X. Suppose given for each U_i a VB-morphism

$$f_i: \alpha|U_i \to \beta|U_i.$$

Each one of the maps $\psi_i f_i$ (defined as in Proposition 7) is a VB-morphism. Furthermore, we can extend $\psi_i f_i$ to a VB-morphism of α into β simply by putting

$$(\psi_i f_i)_x = 0$$

for all $x \notin U_i$. If we now define

$$f: \alpha \to \beta$$

by the formula

$$f_x(v) = \sum \psi_i(x) f_{ix}(v)$$

for all pairs (x, v) with $v \in \alpha_x$, then the sum is actually finite, at each point x, and again by Proposition 7, we see that f is a VB-morphism. We observe that if each f_i is the identity, then $f = \sum \psi_i f_i$ is also the identity.

Proposition 8. *Let X be a manifold admitting partitions of unity. Let $0 \to \alpha \xrightarrow{f} \beta$ be an exact sequence of vector bundles over X. Then there exists a surjective VB-morphism $g: \beta \to \alpha$ whose kernel splits at each point, such that $g \circ f = id$.*

Proof. By the definition of exact sequence, there exists a partition of unity $\{(U_i, \psi_i)\}$ on X such that for each i, we can split the sequence over U_i. In other words, there exists for each i a VB-morphism

$$g_i: \beta|U_i \to \alpha|U_i$$

which is surjective, whose kernel splits, and such that $g_i \circ f_i = id_i$. We let $g = \sum \psi_i g_i$. Then g is a VB-morphism of β into α by what we have just seen, and

$$g \circ f = \sum \psi_i g_i f_i = id.$$

It is trivial that g is surjective because $g \circ f = id$. The kernel of g_x splits at each point x because it has a closed complement, namely $f_x \alpha_x$. This concludes the proof.

If γ is the kernel of β, then we have $\beta \approx \alpha \oplus \gamma$.

A vector bundle π over X will be said to be of **finite type** if there exists a finite trivialisation for π (i.e. a trivialisation $\{(U_i, \tau_i)\}$ such that i ranges over a finite set).

If k is an integer ≥ 1 and $\mathbf{E}$ a topological vector space, then we denote by $\mathbf{E}^k$ the direct product of $\mathbf{E}$ with itself k times.

Proposition 9. *Let X be a manifold admitting partitions of unity. Let π be a vector bundle of finite type in* $\mathrm{VB}(X, \mathbf{E})$, *where $\mathbf{E}$ is a Banach space. Then there exists an integer $k > 0$ and a vector bundle α in* $\mathrm{VB}(X, \mathbf{E}^k)$ *such that $\pi \oplus \alpha$ is trivialisable.*

Proof. We shall prove that there exists an exact sequence

$$0 \to \pi \xrightarrow{\ 1\ } \beta$$

with $E_\beta = X \times \mathbf{E}^k$. Our theorem will follow from the preceding proposition.

Let $\{(U_i, \tau_i)\}$ be a finite trivialisation of π with $i = 1, \ldots, k$. Let $\{(U_i, \psi_i)\}$ be a partition of unity. We define

$$f \colon E_\pi \to X \times \mathbf{E}^k$$

as follows. If $x \in X$ and v is in the fiber of E_π at x, then

$$f_x(v) = (x, \psi_1(x)\tau_1(v), \ldots, \psi_k(x)\tau_k(v)).$$

The expression on the right makes sense, because in case x does not lie in U_i then $\psi_i(x) = 0$ and we do not have to worry about the expression $\tau_i(v)$. If x lies in U_i, then $\tau_i(v)$ means $\tau_{ix}(v)$.

Given any point x, there exists some index i such that $\psi_i(x) > 0$ and hence f is injective. Furthermore, for this x and this index i, f_x maps E_x onto a closed subspace of $\mathbf{E}^k$, which admits a closed complement, namely

$$\mathbf{E} \times \cdots \times 0 \times \cdots \times \mathbf{E}$$

with 0 in the ith place. This proves our proposition.

CHAPTER IV

Vector Fields and Differential Equations

In this chapter, we collect a number of results all of which make use of the notion of differential equation and solutions of differential equations.

Let X be a manifold. A vector field on X assigns to each point x of X a tangent vector, differentiably. (For the precise definition, see §2.) Given x_0 in X, it is then possible to construct a unique curve $\alpha(t)$ starting at x_0 (i.e. such that $\alpha(0) = x_0$) whose derivative at each point is the given vector. It is not always possible to make the curve depend on time t from $-\infty$ to $+\infty$, although it is possible if X is compact.

The structure of these curves presents a fruitful domain of investigation, from a number of points of view. For instance, one may ask for topological properties of the curves, that is those which are invariant under topological automorphisms of the manifold. (Is the curve a closed curve, is it a spiral, is it dense, etc. ?) More generally, following standard procedures, one may ask for properties which are invariant under any given interesting group of automorphisms of X (discrete groups, Lie groups, algebraic groups, Riemannian automorphisms, ad lib.).

We do not go into these theories, each of which proceeds according to its own flavor. We give merely the elementary facts and definitions associated with vector fields, and some simple applications of the existence theorem for their curves.

Throughout this chapter, we assume all manifolds to be Hausdorff, of class C^p with $p \geq 2$ from §2 on, and $p \geq 3$ from §3 on. This latter condition insures that the tangent bundle is of class C^{p-1} with $p - 1 \geq 1$ (or 2).

We shall deal with mappings of several variables, say $f(t, x, y)$, the first of which will be a real variable. We identify $D_1 f(t, x, y)$ with

$$\lim_{h \to 0} \frac{f(t + h, x, y) - f(t, x, y)}{h}.$$

§1. Existence theorem for differential equations

Let $\mathbf{E}$ be a Banach space and U an open subset of $\mathbf{E}$. In this section we consider vector fields locally. The notion will be globalized later, and thus for the moment, we define (the local representation of) a **time-dependent vector field** on U to be a C^p-morphism ($p \geqq 0$)

$$f : J \times U \to \mathbf{E}$$

where J is an open interval containing 0 in $\mathbf{R}$. We think of f as assigning to each point x in U a vector $f(t, x)$ in $\mathbf{E}$, depending on time t.

Let x_0 be a point of U. An **integral curve** for f with **initial condition** x_0 is a mapping of class C^r ($r \geqq 1$)

$$\alpha : J_0 \to U$$

of an open subinterval of J containing 0, into U, such that $\alpha(0) = x_0$ and such that

$$\alpha'(t) = f(t, \alpha(t)).$$

Remark. Let $\alpha : J_0 \to U$ be a continuous map satisfying the condition

$$\alpha(t) = x_0 + \int_0^t f(u, \alpha(u)) \, du.$$

Then α is differentiable, and its derivative is $f(t, \alpha(t))$. Hence α is of class C^1. Furthermore, we can argue recursively, and conclude that if f is of class C^p, then so is α. Conversely, if α is an integral curve for f with initial condition x_0, then it obviously satisfies our integral relation.

Let

$$f : J \times U \to \mathbf{E}$$

be as above, and let x_0 be a point of U. By a **local flow** for f at x_0 we mean a mapping

$$\alpha : J_0 \times U_0 \to U$$

where J_0 is an open subinterval of J containing 0, and U_0 is an open subset of U containing x_0, such that for each x in U_0 the map

$$\alpha_x(t) = \alpha(t, x)$$

is an integral curve for f with initial condition x (i.e. such that $\alpha(0, x) = x$).

As a matter of notation, when we have a mapping with two arguments, say $\varphi(t, x)$, then we denote the separate mappings in each argument when the other is kept fixed by $\varphi_x(t)$ and $\varphi_t(x)$. The choice of letters will always prevent ambiguity.

We shall say that f satisfies a **Lipschitz condition** on U **uniformly with respect to** J if there exists a number $K > 0$ such that

$$|f(t, x) - f(t, y)| \leq K|x - y|$$

for all x, y in U and t in J. We call K a **Lipschitz constant**. If f is of class C^1, it follows at once from the mean value theorem that f is Lipschitz on some open neighborhood $J_0 \times U_0$ of a given point $(0, x_0)$ of U, and that it is bounded on some such neighborhood.

We shall now prove that under a Lipschitz condition, local flows exist and are unique locally. In fact, we prove more, giving a uniformity property for such flows. If b is real > 0, then we denote by J_b the open interval $-b < t < b$.

Proposition 1. *Let J be an open interval of $\mathbf{R}$ containing 0, and U open in the Banach space $\mathbf{E}$. Let x_0 be a point of U, and $a > 0, a < 1$ a real number such that the closed ball $\bar{B}_{3a}(x_0)$ lies in U. Assume that we have a continuous map*

$$f: J \times U \to \mathbf{E}$$

which is bounded by a constant $L \geq 1$ on $J \times U$, and satisfies a Lipschitz condition on U uniformly with respect to J, with constant $K \geq 1$. If $b < a/LK$, then for each x in $\bar{B}_a(x_0)$ there exists a unique flow

$$\alpha: J_b \times B_a(x_0) \to U.$$

If f is of class C^p ($p \geq 1$), then so is each integral curve α_x.

Proof. Let I_b be the closed interval $-b \leq t \leq b$, and let x be a fixed point in $\bar{B}_a(x_0)$. Let M be the set of continuous maps

$$\alpha: I_b \to \bar{B}_{2a}(x_0)$$

of the closed interval into the closed ball of center x_0 and radius $2a$, such that $\alpha(0) = x$. Then M is a complete metric space if we define as usual the distance between maps α, β to be

$$\sup_{t \in I_b} |\alpha(t) - \beta(t)|.$$

We shall now define a mapping

$$S: M \to M$$

of M into itself. For each α in M, we let $S\alpha$ be defined by

$$(S\alpha)(t) = x + \int_0^t f(u, \alpha(u)) \, du.$$

Then $S\alpha$ is certainly continuous, we have $S\alpha(0) = x$, and the distance of any point on $S\alpha$ from x is bounded by the norm of the integral, which is bounded by

$$b \sup |f(u, y)| \le bL < a.$$

Thus $S\alpha$ lies in M.

We contend that our map S is a shrinking map. Indeed,

$$|S\alpha - S\beta| \le b \sup |f(u, \alpha(u)) - f(u, \beta(u))|$$
$$\le bK|\alpha - \beta|,$$

thereby proving our contention.

By the shrinking lemma (Chapter I, §5) our map has a unique fixed point α, and by definition, $\alpha(t)$ satisfies the desired integral relation. Our remark above concludes the proof.

Corollary. *The local flow α in Proposition 1 is continuous. Furthermore, the map $x \mapsto \alpha_x$ of $\bar{B}_a(x_0)$ into the space of curves is continuous, and in fact satisfies a Lipschitz condition.*

Proof. The second statement obviously implies the first. So fix x in $\bar{B}_a(x_0)$ and take y close to x in $\bar{B}_a(x_0)$. We let S_x be the shrinking map of the theorem, corresponding to the initial condition x. Then

$$\|\alpha_x - S_y\alpha_x\| = \|S_x\alpha_x - S_y\alpha_x\| \le |x - y|.$$

Let $C = bK$ so $0 < C < 1$. Then

$$\|\alpha_x - S_y^n\alpha_x\| \le \|\alpha_x - S_y\alpha_x\| + \|S_y\alpha_x - S_y^2\alpha_x\| + \cdots + \|S_y^{n-1}\alpha_x - S_y^n\alpha_x\|$$
$$\le (1 + C + \cdots + C^{n-1})|x - y|.$$

Since the limit of $S_y^n\alpha_x$ is equal to α_y as n goes to infinity, the continuity of the map $x \mapsto \alpha_x$ follows at once. In fact, the map satisfies a Lipschitz condition as stated.

It is easy to formulate a uniqueness theorem for integral curves over their whole domain of definition.

Uniqueness theorem. *Let U be open in $\mathbf{E}$ and let $f: U \to E$ be a vector field of class C^p, $p \ge 1$. Let*

$$\alpha_1: J_1 \to U \qquad and \qquad \alpha_2: J_2 \to U$$

be two integral curves for f with the same initial condition x_0. Then α_1 and α_2 are equal on $J_1 \cap J_2$.

Proof. Let Q be the set of numbers b such that $\alpha_1(t) = \alpha_2(t)$ for

$$0 \leq t < b.$$

Then Q contains some number $b > 0$ by the local uniqueness theorem. If Q is not bounded from above, the equality of $\alpha_1(t)$ and $\alpha_2(t)$ for all $t > 0$ follows at once. If Q is bounded from above, let b be its least upper bound. We must show that b is the right end point of $J_1 \cap J_2$. Suppose that this is not the case. Define curves β_1 and β_2 near 0 by

$$\beta_1(t) = \alpha_1(b + t) \quad \text{and} \quad \beta_2(t) = \alpha_2(b + t).$$

Then β_1 and β_2 are integral curves of f with the initial conditions $\alpha_1(b)$ and $\alpha_2(b)$ respectively. The values $\beta_1(t)$ and $\beta_2(t)$ are equal for small negative t because b is the least upper bound of Q. By continuity it follows that $\alpha_1(b) = \alpha_2(b)$, and finally we see from the local uniqueness theorem that

$$\beta_1(t) = \beta_2(t)$$

for all t in some neighborhood of 0, whence α_1 and α_2 are equal in a neighborhood of b, contradicting the fact that b is a least upper bound of Q. We can argue the same way towards the left end points, and thus prove our statement.

For each $x \in U$, let $J(x)$ be the union of all open intervals containing 0 on which integral curves for f are defined, with initial condition equal to x. The uniqueness statement allows us to define the integral curve uniquely on all of $J(x)$.

Remark. The choice of 0 as the initial time value is made for convenience. From the uniqueness statement one obtains at once (making a time translation) the analogous statement for an integral curve defined on any open interval; in other words, if J_1, J_2 do not necessarily contain 0, and t_0 is a point in $J_1 \cap J_2$ such that $\alpha_1(t_0) = \alpha_2(t_0)$, and also we have the differential equations

$$\alpha_1'(t) = f(\alpha_1(t)) \quad \text{and} \quad \alpha_2'(t) = f(\alpha_2(t)),$$

then α_1 and α_2 are equal on $J_1 \cap J_2$. One can also repeat the proof of Theorem 2 in this case.

In practice, one meets vector fields which may be time dependent, and also depend on parameters. We discuss these to show that their study reduces to the study of the standard case.

Time dependent vector fields

Let J be an open interval, U open in a Banach space $\mathbf{E}$, and

$$f: J \times U \to \mathbf{E}$$

a C^p map, which we view as depending on time $t \in J$. Thus for each t, the map $x \mapsto f(t, x)$ is a vector field on U. Define

$$\bar{f}: J \times U \to \mathbf{R} \times \mathbf{E}$$

by

$$\bar{f}(t, x) = (1, f(t, x)),$$

and view $\bar{f}$ as a time-independent vector field on $J \times U$. Let $\bar{\alpha}$ be its flow, so that

$$\bar{\alpha}'(t, s, x) = \bar{f}(\bar{\alpha}(t, s, x)), \qquad \bar{\alpha}(0, s, x) = (s, x).$$

We note that $\bar{\alpha}$ has its values in $J \times U$ and thus can be expressed in terms of two components. In fact, it follows at once that we can write $\bar{\alpha}$ in the form

$$\bar{\alpha}(t, s, x) = (t + s, \bar{\alpha}_2(t, s, x)).$$

Then $\bar{\alpha}_2$ satisfies the differential equation

$$D_1 \bar{\alpha}_2(t, s, x) = f(t + s, \bar{\alpha}_2(t, s, x))$$

as we see from the definition of $\bar{f}$. Let

$$\beta(t, x) = \bar{\alpha}_2(t, 0, x).$$

Then β is a flow for f, that is β satisfies the differential equation

$$D_1 \beta(t, x) = f(t, \beta(t, x)), \qquad \beta(0, x) = x.$$

Given $x \in U$, any value of t such that α is defined at (t, x) is also such that $\bar{\alpha}$ is defined at $(t, 0, x)$ because α_x and β_x are integral curves of the same vector field, with the same initial condition, hence are equal. Thus the study of time-dependent vector fields is reduced to the study of time-independent ones.

Dependence on parameters

Let V be open in some space $\mathbf{F}$ and let

$$g: J \times V \times U \to \mathbf{E}$$

be a map which we view as a time-dependent vector field on U, also depending on parameters in V. We define

$$G : J \times V \times U \to \mathbf{F} \times \mathbf{E}$$

by

$$G(t, z, y) = \big(0, g(t, z, y)\big)$$

for $t \in J$, $z \in V$, and $y \in U$. This is now a time-dependent vector field on $V \times U$. A local flow for G depends on three variables, say $\beta(t, z, y)$, with initial condition $\beta(0, z, y) = (z, y)$. The map β has two components, and it is immediately clear that we can write

$$\beta(t, z, y) = \big(z, \alpha(t, z, y)\big)$$

for some map α depending on three variables. Consequently α satisfies the differential equation

$$D_1\alpha(t, z, y) = g\big(t, z, \alpha(t, z, y)\big), \qquad \alpha(0, z, y) = y,$$

which gives the flow of our original vector field g depending on the parameters $z \in V$. This procedure reduces the study of differential equations depending on parameters to those which are independent of parameters.

We shall now investigate the behavior of the flow with respect to its second argument, i.e. with respect to the points of U. We shall give two methods for this. The first depends on approximation estimates, and the second on the implicit mapping theorem in function spaces.

Let J_0 be an open subinterval of J containing 0, and let

$$\varphi : J_0 \to U$$

be of class C^1. We shall say that φ is an ε-**approximate solution** of f on J_0 if

$$|\varphi'(t) - f(t, \varphi(t))| \leq \varepsilon$$

for all t in J_0.

Proposition 2. *Let φ_1 and φ_2 be two ε_1- and ε_2-approximate solutions of f on J_0 respectively, and let $\varepsilon = \varepsilon_1 + \varepsilon_2$. Assume that f is Lipschitz with constant K on U uniformly in J_0, or that D_2f exists and is bounded by K on $J \times U$. Let t_0 be a point of J_0. Then for any t in J_0, we have*

$$|\varphi_1(t) - \varphi_2(t)| \leq |\varphi_1(t_0) - \varphi_2(t_0)|e^{K|t - t_0|} + \frac{\varepsilon}{K}\, e^{K|t - t_0|}.$$

Proof. By assumption, we have

$$|\varphi_1'(t) - f(t, \varphi_1(t))| \leq \varepsilon_1$$
$$|\varphi_2'(t) - f(t, \varphi_2(t))| \leq \varepsilon_2.$$

From this we get

$$|\varphi_1'(t) - \varphi_2'(t) + f(t, \varphi_2(t)) - f(t, \varphi_1(t))| \leq \varepsilon.$$

Say $t \geq t_0$ to avoid putting bars around $t - t_0$. Let

$$\psi(t) = |\varphi_1(t) - \varphi_2(t)|$$
$$\omega(t) = |f(t, \varphi_1(t)) - f(t, \varphi_2(t))|.$$

Then, after integrating from t_0 to t, and using triangle inequalities we obtain

$$|\psi(t) - \psi(t_0)| \leq \varepsilon(t - t_0) + \int_{t_0}^{t} \omega(u)\, du$$
$$\leq \varepsilon(t - t_0) + K \int_{t_0}^{t} \psi(u)\, du$$
$$\leq K \int_{t_0}^{t} [\psi(u) + \varepsilon/K]\, du,$$

and finally the recurrence relation

$$\psi(t) \leq \psi(t_0) + K \int_{t_0}^{t} [\psi(u) + \varepsilon/K]\, du.$$

On any closed subinterval of J_0, our map ψ is bounded. If we add ε/K to both sides of this last relation, then we see that our proposition will follow from the next lemma.

Lemma. *Let g be a positive real valued function on an interval, bounded by a number L. Let t_0 be in the interval, say $t_0 \leq t$, and assume that there are numbers $A, K \geq 0$ such that*

$$g(t) \leq A + K \int_{t_0}^{t} g(u)\, du.$$

Then for all integers $n \geq 1$ we have

$$g(t) \leq A \left[1 + \frac{K(t - t_0)}{1!} + \cdots + \frac{K^{n-1}(t - t_0)^{n-1}}{(n - 1)!} \right] + \frac{LK^n(t - t_0)^n}{n!}.$$

Proof. The statement is an assumption for $n = 1$. We proceed by induction. We integrate from t_0 to t, multiply by K, and use the recurrence relation. The statement with $n + 1$ then drops out of the statement with n.

Corollary 1. *Let* $f: J \times U \to \mathbf{E}$ *be continuous, and satisfy a Lipschitz condition on* U *uniformly with respect to* J. *Let* x_0 *be a point of* U. *Then there exists an open subinterval* J_0 *of* J *containing* 0, *and an open subset of* U *containing* x_0 *such that* f *has a unique flow*

$$\alpha: J_0 \times U_0 \to U.$$

We can select J_0 *and* U_0 *such that* α *is continuous and satisfies a Lipschitz condition on* $J_0 \times U_0$.

Proof. Given x, y in U_0 we let $\varphi_1(t) = \alpha(t, x)$ and $\varphi_2(t) = \alpha(t, y)$, using Proposition 1 to get J_0 and U_0. Then $\varepsilon_1 = \varepsilon_2 = 0$. For s, t in J_0 we obtain

$$|\alpha(t, x) - \alpha(s, y)| \leqq |\alpha(t, x) - \alpha(t, y)| + |\alpha(t, y) - \alpha(s, y)|$$

$$\leqq |x - y|e^K + |t - s|L$$

if we take J_0 of small length, and L is a bound for f. Indeed, the term containing $|x - y|$ comes from Proposition 2, and the term containing $|t - s|$ comes from the definition of the integral curve by means of an integral and the bound L for f. This proves our corollary.

Corollary 2. *Let* J *be an open interval of* $\mathbf{R}$ *containing* 0 *and let* U *be open in* $\mathbf{E}$. *Let* $f: J \times U \to \mathbf{E}$ *be a continuous map, which is Lipschitz on* U *uniformly for every compact subinterval of* J. *Let* $t_0 \in J$ *and let* φ_1, φ_2 *be two morphisms of class* C^1 *such that* $\varphi_1(t_0) = \varphi_2(t_0)$ *and satisfying the relation*

$$\varphi'(t) = f(t, \varphi(t))$$

for all t *in* J. *Then* $\varphi_1(t) = \varphi_2(t)$.

Proof. We can take $\varepsilon = 0$ in the proposition.

The above corollary gives us another proof for the uniqueness of integral curves. Given $f: J \times U \to \mathbf{E}$ as in this corollary, we can define an integral curve α for f on a maximal open subinterval of J having a given value $\alpha(t_0)$ for a fixed t_0 in J. Let J be the open interval (a, b) and let (a_0, b_0) be the interval on which α is defined. We want to know when $b_0 = b$ (or $a_0 = a$), that is when the integral curve of f can be continued to the entire interval over which f itself is defined.

There are essentially two reasons why it is possible that the integral curve cannot be extended to the whole domain of definition J, or cannot be extended to infinity in case f is independent of time. One possibility is that the integral curve tends to get out of the open set U, as on the following picture.

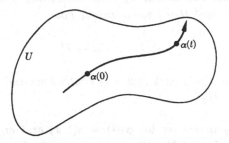

This means that as t approaches b_0, say, the curve $\alpha(t)$ approaches a point which does not lie in U. Such an example can actually be constructed artificially. If we are in a situation when a curve can be extended to infinity, just remove a point from the open set lying on the curve. Then the integral curve on the resulting open set cannot be continued to infinity. The second possibility is that the vector field is unbounded. The next corollary shows that these possibilities are the only ones. In other words, if an integral curve does not tend to get out of the open set, and if the vector field is bounded, then the curve can be continued as far as the original data will allow a priori.

Corollary 3. *Let J be the open interval (a, b) and let U be open in $\mathbf{E}$. Let $f: J \times U \to \mathbf{E}$ be a continuous map, which is Lipschitz on U, uniformly for every compact subset of J. Let α be an integral curve of f, defined on a maximal open subinterval (a_0, b_0) of J. Assume:*

(i) *There exists $\varepsilon > 0$ such that $\overline{\alpha((b_0 - \varepsilon, b_0))}$ is contained in U.*

(ii) *There exists a number $B > 0$ such that $|f(t, \alpha(t))| \leq B$ for all t in $(b_0 - \varepsilon, b_0)$.*

Then $b_0 = b$.

Proof. From the integral expression for α, namely

$$\alpha(t) = \alpha(t_0) + \int_{t_0}^{t} f(u, \alpha(u))\, du,$$

we see that for t_1, t_2 in $(b_0 - \varepsilon, b_0)$ we have

$$|\alpha(t_1) - \alpha(t_2)| \leqq B|t_1 - t_2|.$$

From this it follows that the limit

$$\lim_{t \to b_0} \alpha(t)$$

exists, and is equal to an element x_0 of U (by hypothesis (i)). Assume that $b_0 \neq b$. By the local existence theorem, there exists an integral curve β of f defined on an open interval containing b_0 such that $\beta(b_0) = x_0$ and $\beta'(t) = f(t, \beta(t))$. Then $\beta' = \alpha'$ on an open interval to the left of b_0, and hence α, β differ by a constant on this interval. Since their limit as $t \to b_0$ are equal, this constant is 0. Thus we have extended the domain of definition of α to a larger interval, as was to be shown.

The next proposition describes the solutions of **linear differential equations** depending on parameters.

Proposition 3. *Let J be an open interval of $\mathbf{R}$ containing 0, and let V be an open set in a Banach space. Let $\mathbf{E}$ be a Banach space. Let*

$$g: J \times V \to L(\mathbf{E}, \mathbf{E})$$

be a continuous map. Then there exists a unique map

$$\lambda: J \times V \to L(\mathbf{E}, \mathbf{E})$$

which, for each $x \in V$, is a solution of the differential equation

$$D_1\lambda(t, x) = g(t, x)\lambda(t, x), \qquad \lambda(0, x) = id.$$

This map λ is continuous.

Remark. In the present case of a linear differential equation, it is not necessary to shrink the domain of definition of its flow. Note that the differential equation is on the space of continuous linear maps. The corresponding linear equation on $\mathbf{E}$ itself will come out as a corollary.

Proof of Proposition 3. Let us first fix $x \in V$. Consider the differential equation

$$D_1\lambda(t, x) = g(t, x)\lambda(t, x),$$

with initial condition $\lambda(0, x) = id$. This is a differential equation on $L(\mathbf{E}, \mathbf{E})$, where $f(t, z) = g_x(t)z$ for $z \in L(\mathbf{E}, \mathbf{E})$, and we write $g_x(t)$ instead of $g(t, x)$. Let the notation be as in Corollary 3 of Proposition 2. Then hypothesis (i) is automatically satisfied since the open set U is all of $L(\mathbf{E}, \mathbf{E})$.

On every compact subinterval of J, g_x is bounded, being continuous. Omitting the index x for simplicity, we have

$$\lambda(t) = id + \int_0^t g(u)\lambda(u)\, du,$$

whence for $t \geqq 0$, say,

$$|\lambda(t)| \leqq 1 + B \int_0^t |\lambda(u)|\, du.$$

Using Lemma 1, we see that hypothesis (ii) of Corollary 3 of Proposition 2 is also satisfied. Hence the integral curve is defined on all of J.

We shall now prove the continuity of λ. Let $(t_0, x_0) \in J \times V$. Let I be a compact interval contained in J, and containing t_0 and 0. As a function of t, $\lambda(t, x_0)$ is continuous (even differentiable). Let $C > 0$ be such that $|\lambda(t, x_0)| \leqq C$ for all $t \in I$. Let V_1 be an open neighborhood of x_0 in V such that g is bounded by a constant $K > 0$ on $I \times V_1$.

For $(t, x) \in I \times V_1$ we have

$$|\lambda(t, x) - \lambda(t_0, x_0)| \leqq |\lambda(t, x) - \lambda(t, x_0)| + |\lambda(t, x_0) - \lambda(t_0, x_0)|.$$

The second term on the right is small when t is close to t_0. We investigate the first term on the right, and shall estimate it by viewing $\lambda(t, x)$ and $\lambda(t, x_0)$ as approximate solutions of the differential equation satisfied by $\lambda(t, x)$. We find

$$|D_1\lambda(t, x_0) - g(t, x)\lambda(t, x_0)|$$
$$= |D_1\lambda(t, x_0) - g(t, x)\lambda(t, x_0) + g(t, x_0)\lambda(t, x_0) - g(t, x_0)\lambda(t, x_0)|$$
$$\leqq |g(t, x_0) - g(t, x)|\,|\lambda(t, x_0)| \leqq |g(t, x_0) - g(t, x)|C.$$

By the usual proof of uniform continuity applied to the compact set $I \times \{x_0\}$, given $\varepsilon > 0$, there exists an open neighborhood V_0 of x_0 contained in V_1, such that for all $(t, x) \in I \times V_0$ we have

$$|g(t, x) - g(t, x_0)| < \varepsilon/C.$$

This implies that $\lambda(t, x_0)$ is an ε-approximate solution of the differential equation satisfied by $\lambda(t, x)$. We apply Proposition 2 to the two curves

$$\varphi_0(t) = \lambda(t, x_0) \quad \text{and} \quad \varphi_x(t) = \lambda(t, x)$$

for each $x \in V_0$. We use the fact that $\lambda(0, x) = \lambda(0, x_0) = id$. We then find

$$|\lambda(t, x) - \lambda(t, x_0)| < \varepsilon K_1$$

for some constant $K_1 > 0$, thereby proving the continuity of λ at (t_0, x_0).

Corollary. *Let the notation be as in Proposition 3. For each $x \in V$ and $z \in E$ the curve*

$$\beta(t, x, z) = \lambda(t, x)z$$

with initial condition $\beta(0, x, z) = z$ is a solution of the differential equation

$$D_1\beta(t, x, z) = g(t, x)\beta(t, x, z).$$

Furthermore, β is continuous in its three variables.

Proof. Obvious.

Local smoothness theorem. *Let J be an open interval in $\mathbf{R}$ containing 0 and U open in the Banach space $\mathbf{E}$. Let*

$$f: J \times U \to \mathbf{E}$$

be a C^p-morphism with $p \geq 1$, and let $x_0 \in U$. There exists a unique local flow for f at x_0. We can select an open subinterval J_0 of J containing 0 and an open subset U_0 of U containing x_0 such that the unique local flow

$$\alpha: J_0 \times U_0 \to U$$

is of class C^p, and such that $D_2\alpha$ satisfies the differential equation

$$\boxed{D_1D_2\alpha(t, x) = D_2f\big(t, \alpha(t, x)\big)D_2\alpha(t, x)}$$

on $J_0 \times U_0$ with initial condition $D_2\alpha(0, x) = id$.

Proof. Let

$$g: J \times U \to L(\mathbf{E}, \mathbf{E})$$

be given by $g(t, x) = D_2f(t, \alpha(t, x))$. Select J_1 and U_0 such that α is bounded and Lipschitz on $J_1 \times U_0$ (by Corollary 1 of Proposition 2), and such that g is continuous and bounded on $J_1 \times U_0$. Let J_0 be an open subinterval of J_1 containing 0 such that its closure $\bar{J}_0$ is contained in J_1.

Let $\lambda(t, x)$ be the solution of the differential equation on $L(\mathbf{E}, \mathbf{E})$ given by

$$D_1\lambda(t, x) = g(t, x)\lambda(t, x), \qquad \lambda(0, x) = id,$$

as in Proposition 3. We contend that $D_2\alpha$ exists and is equal to λ on $J_0 \times U_0$. This will prove that $D_2\alpha$ is continuous, on $J_0 \times U_0$.

Fix $x \in U_0$. Let

$$\theta(t, h) = \alpha(t, x + h) - \alpha(t, x).$$

Then

$$D_1\theta(t, h) = D_1\alpha(t, x + h) - D_1\alpha(t, x)$$
$$= f(t, \alpha(t, x + h)) - f(t, \alpha(t, x)).$$

By the mean value theorem, we obtain

$$|D_1\theta(t, h) - g(t, x)\theta(t, h)|$$
$$= |f(t, \alpha(t, x + h)) - f(t, \alpha(t, x)) - D_2f(t, \alpha(t, x))\theta(t, h)|$$
$$\leqq |h| \sup |D_2f(t, y) - D_2f(t, \alpha(t, x))|$$

where y ranges over the segment between $\alpha(t, x)$ and $\alpha(t, x + h)$. By the compactness of $\bar{J}_0$ it follows that our last expression is bounded by $|h|\psi(h)$ where $\psi(h)$ tends to 0 with h, uniformly for t in $\bar{J}_0$. Hence we obtain

$$|\theta'(t, h) - g(t, x)\theta(t, h)| \leqq |h|\psi(h),$$

for all t in $\bar{J}_0$. This shows that $\theta(t, h)$ is an $|h|\psi(h)$ approximate solution for the differential equation satisfied by $\lambda(t, x)h$, namely

$$D_1\lambda(t, x)h - g(t, x)\lambda(t, x)h = 0,$$

with the initial condition $\lambda(0, x)h = h$. We note that $\theta(t, h)$ has the same initial condition, $\theta(0, h) = h$. Taking $t_0 = 0$ in Proposition 2, we obtain the estimate

$$|\theta(t, h) - \lambda(t, x)h| \leqq C_1|h|\psi(h)$$

for all t in $\bar{J}_0$. This proves that $D_2\alpha$ is equal to λ on $J_0 \times U_0$, and is therefore continuous on $J_0 \times U_0$.

We have now proved that $D_1\alpha$ and $D_2\alpha$ exist and are continuous on $J_0 \times U_0$, and hence that α is of class C^1 on $J_0 \times U_0$.

Furthermore, $D_2\alpha$ satisfies the differential equation given in the statement of our theorem on $J_0 \times U_0$. Thus our theorem is proved when $p = 1$.

A flow which satisfies the properties stated in the theorem will be called **locally of class C^p**.

Consider now again the linear equation of Proposition 3. We reformulate it to eliminate formally the parameters, namely we define a vector field

$$G: J \times V \times L(\mathbf{E}, \mathbf{E}) \rightarrow F \times L(\mathbf{E}, \mathbf{E})$$

to be the map such that

$$G(t, x, \omega) = (0, g(t, x)\omega)$$

for $\omega \in L(\mathbf{E}, \mathbf{E})$. The flow for this vector field is then given by the map Λ such that

$$\Lambda(t, x, \omega) = (x, \lambda(t, x)\omega).$$

If g is of class C^1 we can now conclude that the flow Λ is locally of class C^1, and hence putting $\omega = id$, that λ is locally of class C^1.

We apply this to the case when $g(t, x) = D_2 f(t, \alpha(t, x))$, and to the solution $D_2\alpha$ of the differential equation

$$D_1(D_2\alpha)(t, x) = g(t, x)D_2\alpha(t, x)$$

locally at each point $(0, x)$, $x \in U$. Let $p \geq 2$ be an integer and assume our theorem proved up to $p - 1$, so that we can assume α locally of class C^{p-1}, and f of class C^p. Then g is locally of class C^{p-1}, whence $D_2\alpha$ is locally C^{p-1}. From the expression

$$D_1\alpha(t, x) = f(t, \alpha(t, x))$$

we conclude that $D_1\alpha$ is C^{p-1}, whence α is locally C^p.

If f is C^∞, and if we knew that α is of class C^p for every integer p **on its domain of definition**, then we could conclude that α is C^∞; in other words, there is no shrinkage in the inductive application of the local theorem. We shall do this at the end of the section.

We shall now give another proof for the local smoothness of the flow, which depends on a simple application of the implicit mapping theorem in Banach spaces, and was found independently by Pugh and Robbin. One advantage of this proof is that it extends to H^p vector fields, as noted by Ebin and Marsden [7].

Let U be open in $\mathbf{E}$ and let $f: U \to \mathbf{E}$ be a C^p map. Let $b > 0$ and let I_b be the closed interval of radius b centered at 0. Let

$$F = C^0(I_b, \mathbf{E})$$

be the Banach space of continuous maps of I_b into $\mathbf{E}$. We let V be the subset of $\mathbf{F}$ consisting of all continuous curves

$$\sigma : I_b \to U$$

mapping I_b into our open set U. Then it is clear that V is open in $\mathbf{F}$ because for each curve σ the image $\sigma(I_b)$ is compact, hence at a finite distance from the complement of U, so that any curve close to it is also contained in U.

We define a map

$$T : U \times V \to \mathbf{F}$$

by

$$T(x, \sigma) = x + \int_0^t f \circ \sigma - \sigma.$$

Here we omit the dummy variable of integration, and x stands for the constant curve with value x. If we evaluate the curve $T(x, \sigma)$ at t, then by definition we have

$$T(x, \sigma)(t) = x + \int_0^t f(\sigma(u))\, du - \sigma(t).$$

Lemma 1. *The map T is of class C^p, and its second partial derivative is given by the formula*

$$D_2 T(x, \sigma) = \int_0 Df \circ \sigma - I$$

where I is the identity. In terms of t, this reads

$$D_2 T(x, \sigma) h(t) = \int_0^t Df(\sigma(u)) h(u)\, du - h(t).$$

Proof. It is clear that the first partial derivative $D_1 T$ exists and is continuous, in fact C^∞, being linear in x up to a translation. To determine the second partial, we apply the definition of the derivative. The derivative of the map $\sigma \mapsto \sigma$ is of course the identity. We have to get the derivative with respect to σ of the integral expression. We have for small h

$$\left\| \int_0 f \circ (\sigma + h) - \int_0 f \circ \sigma - \int_0 (Df \circ \sigma) h \right\|$$
$$\leq \int_0 |f \circ (\sigma + h) - f \circ \sigma - (Df \circ \sigma) h|.$$

We estimate the expression inside the integral at each point u, with u between 0 and the upper variable of integration. From the mean value theorem, we get

$$|f(\sigma(u) + h(u)) - f(\sigma(u)) - Df(\sigma(u)) h(u)| \leq \|h\| \sup |Df(z_u) - Df(\sigma(u))|$$

where the sup is taken over all points z_u on the segment between $\sigma(u)$ and $\sigma(u) + h(u)$. Since Df is continuous, and using the fact that the image of the curve $\sigma(I_b)$ is compact, we conclude (as in the case of uniform continuity) that as $\|h\| \to 0$, the expression

$$\sup |Df(z_u) - Df(\sigma(u))|$$

also goes to 0. (Put the ε and δ in yourself.) By definition, this gives us the derivative of the integral expression in σ. The derivative of the final term is obviously the identity, so this proves that $D_2 T$ is given by the formula which we wrote down.

This derivative does not depend on x. It is continuous in σ. Namely, we have

$$D_2T(x, \tau) - D_2T(x, \sigma) = \int_0 [Df \circ \tau - Df \circ \sigma].$$

If σ is fixed and τ is close to σ, then $Df \circ \tau - Df \circ \sigma$ is small, as one proves easily from the compactness of $\sigma(I_b)$, as in the proof of uniform continuity. Thus D_2T is continuous. By Proposition 11 of Chapter I, §3, we now conclude that T is of class C^1.

The derivative of D_2T with respect to σ can again be computed as before if Df is itself of class C^1, and thus by induction, if f is of class C^p we conclude that D_2T is of class C^{p-1} so that by the same reference, we conclude that T itself is of class C^p. This proves our lemma.

We observe that a solution of the equation

$$T(x, \sigma) = 0$$

is precisely an integral curve for the vector field, with initial condition equal to x. Thus we are in a situation where we want to apply the implicit mapping theorem.

Lemma 2. *Let $x_0 \in U$. Let $a > 0$ be such that Df is bounded, say by a number $C_1 > 0$, on the ball $B_a(x_0)$ (we can always find such a since Df is continuous at x_0). Let $b < 1/C_1$. Then $D_2T(x, \sigma)$ is invertible for all (x, σ) in $B_a(x_0) \times V$.*

Proof. We have an estimate

$$\left| \int_0^t Df(\sigma(u))h(u)\, du \right| \leq bC_1 \|h\|.$$

This means that

$$|D_2T(x, \sigma) + I| < 1,$$

and hence that $D_2T(x, \sigma)$ is invertible, as a continuous linear map, thus proving Lemma 2. We are ready to reprove the local smoothness theorem by the present means, when p is an integer, namely:

Let p be a positive integer, and let $f \colon U \to \mathbf{E}$ be a C^p vector field. Let $x_0 \in U$. Then there exist numbers $a, b > 0$ such that the local flow

$$\alpha \colon J_b \times B_a(x_0) \to U$$

is of class C^p.

Proof. We take a so small and then b so small that the local flow exists and is uniquely determined by Proposition 1. We then take b smaller and a smaller so as to satisfy the hypotheses of Lemma 2. We can then apply the implicit mapping theorem to conclude that the map $x \mapsto \alpha_x$ is of class C^p. Of course, we have to consider the flow α and still must show that α itself is of class C^p. It will suffice to prove that $D_1\alpha$ and $D_2\alpha$ are of class C^{p-1}, by Proposition 11 of Chapter I, §3. We first consider the case $p = 1$.

We could derive the continuity of α from the Corollary of Proposition 1, but we can also get it as an immediate consequence of the continuity of the map $x \mapsto \alpha_x$. Indeed, fixing (s, y) we have

$$|\alpha(t, x) - \alpha(s, y)| \leq |\alpha(t, x) - \alpha(t, y)| + |\alpha(t, y) - \alpha(s, y)|$$

$$\leq \|\alpha_x - \alpha_y\| + |\alpha_y(t) - \alpha_y(s)|.$$

Since α_y is continuous (being differentiable), we get the continuity of α. Since

$$D_1\alpha(t, x) = f(\alpha(t, x)),$$

we conclude that $D_1\alpha$ is a composite of continuous maps, whence continuous.

Let φ be the derivative of the map $x \mapsto \alpha_x$, so that

$$\varphi : B_a(x_0) \to L(\mathbf{E}, C^0(I_b, \mathbf{E})) = L(\mathbf{E}, \mathbf{F})$$

is of class C^{p-1}. Then

$$\alpha_{x+w} - \alpha_x = \varphi(x)w + |w|\psi(w)$$

where $\psi(w) \to 0$ as $w \to 0$. Evaluating at t, we find

$$\alpha(t, x + w) - \alpha(t, x) = (\varphi(x)w)(t) + |w|\psi(w)(t),$$

and from this we see that

$$D_2\alpha(t, x)w = (\varphi(x)w)(t).$$

Then

$$|D_2\alpha(t, x)w - D_2\alpha(s, y)w|$$

$$\leq |(\varphi(x)w)(t) - (\varphi(y)w)(t)| + |(\varphi(y)w)(t) - (\varphi(y)w)(s)|.$$

The first term on the right is bounded by

$$|\varphi(x) - \varphi(y)| \, |w|$$

so that

$$|D_2\alpha(t, x) - D_2\alpha(t, y)| \leq |\varphi(x) - \varphi(y)|.$$

The curve $\varphi(y)w$ is continuous, so that if t is close to s, the second term on the right is small. This proves the continuity of $D_2\alpha$, and concludes the proof that α is of class C^1.

We have

$$\alpha(t, x) = x + \int_0^t f(\alpha(u, x)) \, du.$$

We can differentiate under the integral sign with respect to the parameter x and thus obtain

$$D_2\alpha(t, x) = I + \int_0^t Df(\alpha(u, x))D_2\alpha(u, x) \, du,$$

where I is a constant linear map (the identity). Differentiating with respect to t yields the linear differential equation satisfied by $D_2\alpha$, namely

$$D_1D_2\alpha(t, x) = Df(\alpha(t, x))D_2\alpha(t, x)$$

and this differential equation depends on time and parameters. We have seen earlier how such equations can be reduced to the ordinary case. We now conclude that locally, by induction, $D_2\alpha$ is of class C^{p-1} since Df is of class C^{p-1}. Since

$$D_1\alpha(t, x) = f(\alpha(t, x)),$$

we conclude by induction that $D_1\alpha$ is C^{p-1}. Hence α is of class C^p by Proposition 11 of Chapter I, §3. Note that each time we use induction, the domain of the flow may shrink. We have proved the theorem when p is an integer.

We now give the arguments needed to globalize the smoothness. We may limit ourselves to the time-independent case. We have seen that the time-dependent case reduces to the other.

Let U be open in a Banach space $\mathbf{E}$, and let $f: U \to \mathbf{E}$ be a C^p vector field. We let $J(x)$ be the domain of the integral curve with initial condition equal to w.

Let $\mathfrak{D}(f)$ be the set of all points (t, x) in $\mathbf{R} \times U$ such that t lies in $J(x)$. Then we have a map

$$\alpha: \mathfrak{D}(f) \to U$$

defined on all of $\mathcal{D}(f)$, letting $\alpha(t, x) = \alpha_x(t)$ be the integral curve on $J(x)$ having x as initial condition. We call this the **flow** determined by f, and we call $\mathcal{D}(f)$ its **domain of definition**.

Lemma 3. *Let $f: U \to E$ be a C^p vector field on the open set U of E, and let α be its flow. Abbreviate $\alpha(t, x)$ by tx, if (t, x) is in the domain of definition of the flow. Let $x \in U$. If t_0 lies in $J(x)$, then*

$$J(t_0 x) = J(x) - t_0$$

(translation of $J(x)$ by $-t_0$), and we have for all t in $J(x) - t_0$:

$$t(t_0 x) = (t + t_0)x.$$

Proof. The two curves defined by

$$t \mapsto \alpha\big(t, \alpha(t_0, x)\big) \qquad \text{and} \qquad t \mapsto \alpha(t + t_0, x)$$

are integral curves of the same vector field, with the same initial condition $t_0 x$ at $t = 0$. Hence they have the same domain of definition $J(t_0 x)$. Hence t_1 lies in $J(t_0 x)$ if and only if $t_1 + t_0$ lies in $J(x)$. This proves the first assertion. The second assertion comes from the uniqueness of the integral curve having given initial condition, whence the theorem follows.

Theorem 1 (Global smoothness of the flow). *If f is of class C^p (with $p \leq \infty$), then its flow is of class C^p on its domain of definition.*

Proof. First let p be an integer ≥ 1. We know that the flow is locally of class C^p at each point $(0, x)$, by the local theorem. Let $x_0 \in U$ and let $J(x_0)$ be the maximal interval of definition of the integral curve having x_0 as initial condition. Let $\mathcal{D}(f)$ be the domain of definition of the flow, and let α be the flow. Let Q be the set of numbers $b > 0$ such that for each t with $0 \leq t < b$ there exists an open interval J containing t and an open set V containing x_0 such that $J \times V$ is contained in $\mathcal{D}(f)$ and such that α is of class C^p on $J \times V$. Then Q is not empty by the local theorem. If Q is not bounded from above, then we are done looking toward the right end point of $J(x_0)$. If Q is bounded from above, we let b be its least upper bound. We must prove that b is the right end point of $J(x_0)$. Suppose that this is not the case. Then $\alpha(b, x_0)$ is defined. Let $x_1 = \alpha(b, x_0)$. By the local theorem, we have a unique local flow at x_1, which we denote by β:

$$\beta: J_a \times B_a(x_1) \to U, \qquad\qquad \beta(0, x) = x,$$

defined for some open interval $J_a = (-a, a)$ and open ball $B_a(x_1)$ of radius a centered at x_1. Let δ be so small that whenever $b - \delta < t < b$ we have

$$\alpha(t, x_0) \in B_{a/4}(x_1).$$

We can find such δ because

$$\lim_{t \to b} \alpha(t, x_0) = x_1$$

by continuity. Select a point t_1 such that $b - \delta < t_1 < b$. By the hypothesis on b, we can select an open interval J_1 containing t_1 and an open set U_1 containing x_0 so that

$$\alpha : J_1 \times U_1 \to B_{a/2}(x_1)$$

maps $J_1 \times U_1$ into $B_{a/2}(x_1)$. We can do this because α is continuous at (t_1, x_0), being in fact C^p at this point. If $|t - t_1| < a$ and $x \in U_1$, we define

$$\varphi(t, x) = \beta(t - t_1, \alpha(t_1, x)).$$

Then

$$\varphi(t_1, x) = \beta(0, \alpha(t_1, x)) = \alpha(t_1, x)$$

and

$$D_1\varphi(t, x) = D_1\beta(t - t_1, \alpha(t_1, x))$$
$$= f(\beta(t - t_1, \alpha(t_1, x)))$$
$$= f(\varphi(t, x)).$$

Hence both φ_x and α_x are integral curves for f with the same value at t_1. They coincide on any interval on which they are defined by the uniqueness theorem. If we take δ very small compared to a, say $\delta < a/4$, we see that φ is an extension of α to an open set containing (t_1, x_0), and also containing (b, x_0). Furthermore, φ is of class C^p, thus contradicting the fact that b is strictly smaller than the end point of $J(x_0)$. Similarly, one proves the analogous statement on the other side, and we therefore see that $\mathfrak{D}(f)$ is open in $\mathbf{R} \times U$ and that α is of class C^p on $\mathfrak{D}(f)$, as was to be shown.

The idea of the above proof is very simple geometrically. We go as far to the right as possible in such a way that the given flow α is of class C^p locally at (t, x_0). At the point $\alpha(b, x_0)$ we then use the flow β to extend differentiably the flow α in case b is not the right-hand point of $J(x_0)$. The flow β at $\alpha(b, x_0)$ has a fixed local domain of definition, and we simply take t close enough to b so that β gives an extension of α, as described in the above proof.

Of course, if f is of class C^∞, then we have shown that α is of class C^p for each positive integer p, and therefore the flow is also of class C^∞.

In the next section, we shall see how these arguments globalize even more to manifolds.

§2. Vector fields, curves, and flows

Let X be a manifold of class C^p with $p \geq 2$. We recall that X is assumed to be Hausdorff. Let $\pi : T(X) \to X$ be its tangent bundle. Then $T(X)$ is of class C^{p-1}, $p \geq 1$.

By a (time-independent) **vector field** on X we mean a cross section of the tangent bundle, i.e. a morphism (of class C^{p-1})

$$\xi : X \to T(X)$$

such that $\xi(x)$ lies in the tangent space $T_x(X)$ for each $x \in X$, or in other words, such that $\pi\xi = id$.

If $T(X)$ is trivial, and say X is an **E**-manifold, so that we have a VB-isomorphism of $T(X)$ with $X \times \mathbf{E}$, then the morphism ξ is completely determined by its projection on the second factor, and we are essentially in the situation of the preceding paragraph, except for the fact that our vector field is independent of time. In such a product representation, the projection of ξ on the second factor will be called the **local representation** of ξ. It is a C^{p-1}-morphism

$$f : X \to \mathbf{E}$$

and $\xi(x) = \big(x, f(x)\big)$. We shall also say that ξ is **represented by f locally** if we work over an open subset U of X over which the tangent bundle admits a trivialisation. We then frequently use ξ itself to denote this local representation.

Let J be an open interval of $\mathbf{R}$. The tangent bundle of J is then $J \times \mathbf{R}$ and we have a canonical section ι such that $\iota(t) = 1$ for all $t \in J$. We sometimes write ι_t instead of $\iota(t)$.

By a **curve** in X we mean a morphism (always of class ≥ 1 unless otherwise specified)

$$\alpha : J \to X$$

from an open interval in $\mathbf{R}$ into X. If $g: X \to Y$ is a morphism, then $g \circ \alpha$ is a curve in Y. From a given curve α, we get an induced map on the tangent bundles:

$$
\begin{array}{ccc}
J \times \mathbf{R} & \xrightarrow{\ \alpha_*\ } & T(X) \\
\downarrow & & \downarrow{\scriptstyle \pi} \\
J & \xrightarrow{\ \alpha\ } & X
\end{array}
$$

and $\alpha_* \circ \imath$ will be denoted by α' or by $d\alpha/dt$ if we take its value at a point t in J. Thus α' is a curve in $T(X)$, of class C^{p-1} if α is of class C^p. Unless otherwise specified, it is always understood in the sequel that we start with enough differentiability to begin with so that we never end up with maps of class < 1. Thus to be able to take derivatives freely we have to take X and α of class C^p with $p \geq 2$.

If $g: X \to Y$ is a morphism, then

$$(g \circ \alpha)'(t) = g_* \alpha'(t).$$

This follows at once from the functoriality of the tangent bundle and the definitions.

Suppose that J contains 0, and let us consider curves defined on J and such that $\alpha(0)$ is equal to a fixed point x_0. We could say that two such curves α_1, α_2 are **tangent** at 0 if $\alpha_1'(0) = \alpha_2'(0)$. The reader will verify immediately that there is a natural bijection between tangency classes of curves with $\alpha(0) = x_0$ and the tangent space $T_{x_0}(X)$ of X at x_0. The tangent space could therefore have been defined alternatively by taking equivalence classes of curves through the point.

Let ξ be a vector field on X and x_0 a point of X. An **integral curve** for the vector field ξ with **initial condition** x_0, or starting at x_0, is a curve (of class C^{p-1})

$$\alpha: J \to X$$

mapping an open interval J of $\mathbf{R}$ containing 0 into X, such that $\alpha(0) = x_0$ and such that

$$\alpha'(t) = \xi(\alpha(t))$$

for all $t \in J$. Using a local representation of the vector field, we know from the preceding section that integral curves exist locally. The next theorem gives us their global existence and uniqueness.

Theorem 2. *Let $\alpha_1: J_1 \to X$ and $\alpha_2: J_2 \to X$ be two integral curves of the vector field ξ on X, with the same initial condition x_0. Then α_1 and α_2 are equal on $J_1 \cap J_2$.*

Proof. Let J^* be the set of points t such that $\alpha_1(t) = \alpha_2(t)$. Then J^* certainly contains a neighborhood of 0 by the local uniqueness theorem. Furthermore, since X is Hausdorff, we see that J^* is closed. We must show that it is open. Let t^* be in J^* and define β_1, β_2 near 0 by

$$\beta_1(t) = \alpha_1(t^* + t)$$
$$\beta_2(t) = \alpha_2(t^* + t).$$

Then β_1 and β_2 are integral curves of ζ with initial condition $\alpha_1(t^*)$ and $\alpha_2(t^*)$ respectively, so by the local uniqueness theorem, β_1 and β_2 agree in a neighborhood of 0 and thus α_1, α_2 agree in a neighborhood of t^*, thereby proving our theorem.

It follows from Theorem 2 that the union of the domains of all integral curves of ζ with a given initial condition x_0 is an open interval which we denote by $J(x_0)$. Its end points are denoted by $t^+(x_0)$ and $t^-(x_0)$ respectively. (We do not exclude $+\infty$ and $-\infty$.)

Let $\mathfrak{D}(\zeta)$ be the subset of $\mathbf{R} \times X$ consisting of all points (t, x) such that

$$t^-(x) < t < t^+(x).$$

A (global) **flow** for ζ is a mapping

$$\alpha : \mathfrak{D}(\zeta) \to X$$

such that for each $x \in X$, the map $\alpha_x : J(x) \to X$ given by

$$\alpha_x(t) = \alpha(t, x)$$

defined on the open interval $J(x)$ is a morphism and is an integral curve for ζ with initial condition x. When we select a chart at a point x_0 of X, then one sees at once that this definition of flow coincides with the definition we gave locally in the previous section, for the local representation of our vector field.

Given a point $x \in X$ and a number t, we say that tx is **defined** if (t, x) is in the domain of α, and we denote $\alpha(t, x)$ by tx in that case.

Theorem 3. *Let ζ be a vector field on X, and α its flow. Let x be a point of X. If t_0 lies in $J(x)$, then*

$$J(t_0x) = J(x) - t_0$$

(translation of $J(x)$ by $-t_0$), and we have for all t in $J(x) - t_0$:

$$t(t_0x) = (t + t_0)x.$$

Proof. Our first assertion follows immediately from the maximality assumption concerning the domains of the integral curves. The second is equivalent to saying that the two curves given by the left-hand side and right-hand side of the last equality are equal. They are both integral curves for the vector field, with initial condition $t_0 x$ and must therefore be equal.

In particular, if t_1, t_2 are two numbers such that $t_1 x$ is defined and $t_2(t_1 x)$ is also defined, then so is $(t_1 + t_2)x$ and they are equal.

Theorem 4. *Let ξ be a vector field on X, and x a point of X. Assume that $t^+(x) < \infty$. Given a compact set $A \subset X$, there exists $\varepsilon > 0$ such that for all $t > t^+(x) - \varepsilon$, the point tx does not lie in A, and similarly for t^-.*

Proof. Suppose such ε does not exist. Then we can find a sequence t_n of real numbers approaching $t({}^+x)$ from below, such that $t_n x$ lies in A. Since A is compact, taking a subsequence if necessary, we may assume that $t_n x$ converges to a point in A. By the local existence theorem, there exists a neighborhood U of this point y and a number $\delta > 0$ such that $t^+(z) > \delta$ for all $z \in U$. Taking n large, we have

$$t^+(x) < \delta + t_n$$

and $t_n x$ is in U. Then by Theorem 3,

$$t^+(x) = t^+(t_n x) + t_n > \delta + t_n > t({}^+x)$$

contradiction.

Corollary. *If X is compact, and ξ is a vector field on X, then $\mathfrak{D}(\xi) = \mathbf{R} \times X$.*

It is also useful to give one other criterion when $\mathfrak{D}(\xi) = \mathbf{R} \times X$, even when X is not compact. Such a criterion must involve some structure stronger than the differentiable structure (essentially a metric of some sort), because we can always dig holes in a compact manifold by taking away a point.

Proposition 4. *Let $\mathbf{E}$ be a Banach space, and X an $\mathbf{E}$-manifold. Let ξ be a vector field on X. Assume that there exist numbers $a > 0$ and $K > 0$ such that every point x of X admits a chart (U, φ) at x such that the local representation f of the vector field on this chart is bounded by K, and so is its derivative f'. Assume also that φU contains a ball of radius a around φx. Then $\mathfrak{D}(\xi) = \mathbf{R} \times X$.*

Proof. This follows at once from the global continuation theorem, and the uniformity of Proposition 1, §1.

We shall prove finally that $\mathfrak{D}(\xi)$ is open and that α is a morphism.

Theorem 5. *Let ξ be a vector field of class C^{p-1} on the C^p-manifold X $(2 \leq p \leq \infty)$. Then $\mathfrak{D}(\xi)$ is open in $\mathbf{R} \times X$, and the flow α for ξ is a C^{p-1}-morphism.*

Proof. Let first p be an integer ≥ 2. Let $x_0 \in X$. Let J^* be the set of points in $J(x_0)$ for which there exists a number $b > 0$ and an open neighborhood U of x_0 such that $(t - b, t + b)\, U$ is contained in $\mathfrak{D}(\xi)$, and such that the restriction of the flow α to this product is a C^{p-1}-morphism. Then J^* is open in $J(x_0)$, and certainly contains 0 by the local theorem. We must therefore show that J^* is closed in $J(x_0)$.

Let s be in its closure. By the local theorem, we can select a neighborhood V of $sx_0 = \alpha(s, x_0)$ so that we have a unique local flow

$$\beta: J_a \times V \to X$$

for some number $a > 0$, with initial condition $\beta(0, x) = x$ for all $x \in V$, and such that this local flow β is C^{p-1}.

The integral curve with initial condition x_0 is certainly continuous on $J(x_0)$. Thus tx_0 approaches sx_0 as t approaches s. Let V_1 be a given small neighborhood of sx_0 contained in V. By the definition of J^*, we can find an element t_1 in J^* very close to s, and a small number b (compared to a) and a small neighborhood U of x_0 such that α maps the product

$$(t_1 - b, t_1 + b) \times U$$

into V_1, and is C^{p-1} on this product. For $t \in J_a + t_1$ and $x \in U$, we define

$$\varphi(t, x) = \beta(t - t_1, \alpha(t_1, x)).$$

Then $\varphi(t_1, x) = \beta(0, \alpha(t_1, x)) = \alpha(t_1, x)$, and

$$D_1\varphi(t, x) = D_1\beta(t - t_1, \alpha(t_1, x))$$

$$= \xi(\beta(t - t_1, \alpha(t_1, x))$$

$$= \xi(\varphi(t, x)).$$

Hence both φ_x, α_x are integral curves for ξ, with the same value at t_1. They coincide on any interval on which they are defined, so that φ_x is a continuation of α_x to a bigger interval containing s. Since α is C^{p-1} on the product $(t_1 - b, t_1 + b) \times U$, we conclude that φ is also C^{p-1} on $(J_a + t_1) \times U$.

From this we see that $\mathfrak{D}(\xi)$ is open in $\mathbf{R} \times X$, and that α is of class C^{p-1} on its full domain $\mathfrak{D}(\xi)$. If $p = \infty$, then we can now conclude that α is of class C^r for each positive integer r on $\mathfrak{D}(\xi)$, and hence is C^∞, as desired.

Corollary 1. *For each $t \in \mathbf{R}$, the set of $x \in X$ such that (t, x) is contained in the domain $\mathfrak{D}(\xi)$ is open in X.*

Corollary 2. *The functions $t^+(x)$ and $t^-(x)$ are upper and lower semi-continuous respectively.*

Theorem 6. *Let ξ be a vector field on X and α its flow. Let $\mathfrak{D}_t(\xi)$ be the set of points x of X such that (t, x) lies in $\mathfrak{D}(\xi)$. Then $\mathfrak{D}_t(\xi)$ is open for each $t \in \mathbf{R}$, and α_t is an isomorphism of $\mathfrak{D}_t(\xi)$ onto an open subset of X. In fact, $\alpha_t(\mathfrak{D}_t) = \mathfrak{D}_{-t}$ and $\alpha_t^{-1} = \alpha_{-t}$.*

Proof. Immediate from the preceding theorem.

Corollary. *If x_0 is a point of X and t is in $J(x_0)$, then there exists an open neighborhood U of x_0 such that t lies in $J(x)$ for all $x \in U$, and the map*

$$x \mapsto tx$$

is an isomorphism of U onto an open neighborhood of tx_0.

Critical points

Let ξ be a vector field. A **critical point** of ξ is a point x_0 such that $\xi(x_0) = 0$. Critical points play a significant role in the study of vector fields, notably in the Morse theory. We don't go into this here, but just make a few remarks to show at the basic level how they affect the behavior of integral curves.

Remark 1. *If α is an integral curve of a C^1 vector field, ξ, and α passes through a critical point, then α is constant, that is $\alpha(t) = x_0$ for all t.*

Proof. The constant curve through x_0 is an integral curve for the vector field, and the uniqueness theorem shows that it is the only one.

Some smoothness of the vector field in addition to continuity must be assumed for the uniqueness. For instance, the following picture illustrates a situation where the integral curves are not unique. They consist in translations of the curve $y = x^3$ in the plane. The vector field is continuous but not locally Lipschitz.

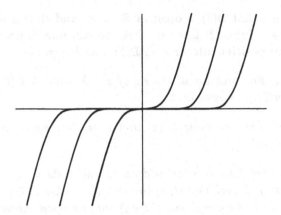

Remark 2. *Let ξ be a vector field and α an integral curve for ξ. Assume that all $t \geq 0$ are in the domain of α, and that*

$$\lim_{t \to 0} \alpha(t) = x_1$$

exists. Then x_1 is a critical point for ξ, that is $\xi(x_1) = 0$.

Proof. Selecting t large, we may assume that we are dealing with the local representation f of the vector field near x_1. Then for $t' > t$ large, we have

$$\alpha(t') - \alpha(t) = \int_t^{t'} f(\alpha(u)) \, du.$$

Write $f(\alpha(u)) = f(x_1) + g(u)$, where $\lim g(u) = 0$. Then

$$|f(x_1)| \, |t' - t| \leq |\alpha(t') - \alpha(t)| + |t' - t| \sup |g(u)|,$$

where the sup is taken for u large, and hence for small values of $g(u)$. Dividing by $|t' - t|$ shows that $f(x_1)$ is arbitrarily small, hence equal to 0, as was to be shown.

Remark 3. *Suppose on the other hand that x_0 is not a critical point of the vector field ξ. Then there exists a chart at x_0 such that the local representation of the vector field on this chart is constant.*

Proof. In an arbitrary chart the vector field has a representation as a morphism

$$\xi : U \to E$$

near x_0. Let α be its flow. We wish to "straighten out" the integral curves of the vector field according to the next figure.

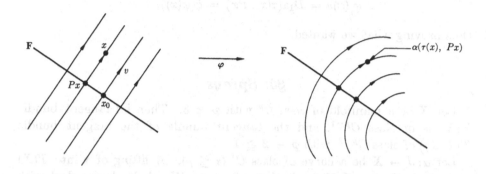

In other words, let $v = \xi(x_0)$. We want to find a local isomorphism φ at x_0 such that

$$\varphi'(x)v = \xi(\varphi(x)).$$

We inspire ourselves from the picture. Without loss of generality, we may assume that $x_0 = 0$. Let λ be a functional such that $\lambda(v) \neq 0$. We decompose $\mathbf{E}$ as a direct sum

$$\mathbf{E} = \mathbf{F} \oplus \mathbf{R}v,$$

where $\mathbf{F}$ is the kernel of λ. Let P be the projection on $\mathbf{F}$. We can write any x near 0 in the form

$$x = Px + \tau(x)v,$$

where

$$\tau(x) = \frac{\lambda(x)}{\lambda(v)}.$$

We then bend the picture on the left to give the picture on the right using the flow α of ξ, namely we define

$$\varphi(x) = \alpha(\tau(x), Px).$$

This means that starting at Px, instead of going linearly in the direction of v for a time $\tau(x)$, we follow the flow (integral curve) for this amount of time. We find that

$$\varphi'(x) = D_1\alpha(\tau(x), Px)\frac{\lambda}{\lambda(v)} + D_2\alpha(\tau(x), Px)P.$$

Hence $\varphi'(0) = id$, so by the inverse mapping theorem, φ is a local isomorphism at 0. Furthermore, since $Pv = 0$ by definition, we have

$$\varphi'(x)v = D_1\alpha(\tau(x), Px) = \xi(\varphi(x)),$$

thus proving what we wanted.

§3. Sprays

Let X be a manifold of class C^p with $p \geq 3$. Then its tangent bundle $T(X)$ is of class C^{p-1}, and the tangent bundle of the tangent bundle $T(T(X))$ of class C^{p-2}, with $p - 2 \geq 1$.

Let $\alpha: J \to X$ be a curve of class C^s ($s \leq p$). A **lifting** of α into $T(X)$ is a curve $\beta: J \to T(X)$ such that $\pi\beta = \alpha$. We shall always deal with $s \geq 2$ so that a lift will be assumed of class $s - 1 \geq 1$. Such lifts always exist, for instance the curve α' discussed in the previous section, called the **canonical lifting** of α.

A **second order differential equation** over X is a vector field ξ on the tangent bundle $T(X)$ (of class C^{p-1}) such that, if π denotes the canonical projection of $T(X)$ on X, then

$$\pi_*\xi(v) = v$$

for all v in $T(X)$. Observe that the succession of symbols makes sense, because

$$\pi_*: TT(X) \to T(X)$$

maps the double tangent bundle into $T(X)$ itself. It follows immediately from the definitions that ξ is a second order differential equation if and only if it satisfies the following condition: Each integral curve β of ξ is equal to the canonical lifting of $\pi\beta$, in other words

$$(\pi\beta)' = \beta.$$

Here, $\pi\beta$ is the canonical projection of β on X, and if we put the argument t, then our formula reads

$$(\pi\beta)'(t) = \beta(t)$$

for all t in the domain of β. We shall give an example later.

We shall be interested in special kinds of second order differential equations. Before we discuss these, we make a few technical remarks.

Let s be a real number, and $\pi: E \to X$ is a vector bundle. If v is in E, so in E_x for some x in X, then sv is again in E_x since E_x is a vector space. We shall identify s with the mapping of E into itself given by this scalar multiplication. This mapping is in fact a VB-morphism, and even a VB-isomorphism if $s \neq 0$. Then

$$s_* \colon T(E) \to T(E)$$

is the usual induced map on the tangent bundle of E.

Now let $E = T(X)$ be the tangent bundle itself. Then our map s satisfies the following properties. First,

$$s_* s = s s_*,$$

which follows from the linearity of s_* on each fiber. Second, if $\alpha \colon J \to X$ is a curve, and α_1 is the curve defined by $\alpha_1(t) = \alpha(st)$, then

$$\alpha_1'(t) = s\alpha'(st),$$

this being the chain rule for differentiation.

Let ξ be a second order differential equation on X. If v is a vector in $T(X)$, let β_v be the unique integral curve of ξ with initial condition v (i.e. such that $\beta_v(0) = v$). Let $\mathfrak{D}$ be the set of vectors on $T(X)$ such that β_v is defined at least on the interval $[0, 1]$. We know from Corollary 1 of Theorem 5, §2, that $\mathfrak{D}$ is an open set in $T(X)$, and by the theorem itself, the map

$$v \mapsto \beta_v(1)$$

is a morphism of $\mathfrak{D}$ into $T(X)$. We now define the **exponential** map

$$\exp \colon \mathfrak{D} \to X$$

to be

$$\exp(v) = \pi \beta_v(1).$$

Then exp is a C^{p-2}-morphism. We also call $\mathfrak{D}$ the **domain of the exponential map** (associated with ξ).

Let X be a manifold of class C^p ($p \geq 3$) and let ξ be a second order differential equation on X. Then the following conditions are equivalent, and define what we shall call a **spray** over X. (In the first three conditions, the sentence should begin with "for each v in $T(X)$.")

SPR 1 *A number t is in the domain of β_v if and only if 1 is in the domain of β_{tv}, and then*

$$\pi \beta_v(t) = \pi \beta_{tv}(1).$$

SPR 2 *If s, t are numbers, st is in the domain of β_v if and only if s is in the domain of β_{tv}, and then*

$$\pi\beta_{tv}(s) = \pi\beta_v(st).$$

SPR 3. *A number t is in the domain of β_{sv} if and only if st is in the domain of β_v and then*

$$\beta_{sv}(t) = s\beta_v(st).$$

SPR 4. *For all $s \in \mathbf{R}$ and $v \in T(X)$, we have*

$$\xi(sv) = s_* s\xi(v).$$

We shall now prove that all these conditions are equivalent.

Assume **SPR 1**. Then st is in the domain of β_r if and only if 1 is in the domain of β_{stv}, and s is in the domain of β_{tv} if and only if 1 is in the domain of β_{stv}. This proves the first assertion of **SPR 2**, and again by **SPR 1**, assuming these relations, we get what we want.

It is similarly clear that **SPR 2** implies **SPR 1**.

Assume **SPR 2**. To prove **SPR 3**, we have

$$(\pi\beta_{sv})'(t) = \frac{d}{dt}\,\pi\beta_v(st) = s(\pi\beta_v)'(st) = s\beta_v(st).$$

Using the fact that $(\pi\beta_{sv})' = \beta_{sv}$, we get what we want. It is obvious that **SPR 3** implies **SPR 2**.

Assume **SPR 3**. Since β_v is an integral curve of ξ for each v, with initial condition v, we have by definition

$$\beta'_{sv}(0) = \xi(sv).$$

Using our assumption, we also have

$$\beta'_{sv}(t) = \frac{d}{dt}\left(s\beta_v(st)\right) = s_* s\beta'_v(st).$$

Putting $t = 0$ and using the assumption once more gives us the condition **SPR 4**.

Finally, assume **SPR 4**, and let s be fixed. For all t such that st is in the domain of β_v, the curve $\beta_v(st)$ is defined and we have

$$\frac{d}{dt}\left(s\beta_v(st)\right) = s_* s\beta'_v(st) = s_* s\xi\big(\beta_v(st)\big) = \xi\big(s\beta_v(st)\big).$$

Hence the curve $s\beta_v(st)$ is an integral curve for ζ, with initial condition $s\beta_v(0) = sv$. By uniqueness we must have

$$s\beta_v(st) = \beta_{sv}(t).$$

This proves **SPR 3**. We are done.

To summarize, a spray is a vector field on $T(X)$ which satisfies the following two conditions:

$$\pi_* \xi(v) = v,$$

$$\xi(sv) = s_* s\xi(v).$$

The second condition is equivalent with any one of the first three **SPR** conditions, and is just **SPR 4**. It is particularly useful to give examples of sprays. Indeed, it is clear from these two, that **sprays form a convex set**! Hence if we can exhibit sprays over open subsets of Banach spaces, then we can glue them together by means of partitions of unity, and we obtain at once the following global existence theorem.

Theorem 7. *Let X be a manifold of class C^p ($p \geqq 3$). If X admits partitions of unity, then there exists a spray over X.*

We give an example of a spray locally.

Example. Let U be open in the Banach space $\mathbf{E}$, so that $T(U) = U \times \mathbf{E}$, and $T(T(U)) = (U \times \mathbf{E}) \times (\mathbf{E} \times \mathbf{E})$. Then $\pi: U \times \mathbf{E} \to U$ is simply the projection, and we have a commutative diagram:

$$
\begin{array}{ccc}
(U \times \mathbf{E}) \times (\mathbf{E} \times \mathbf{E}) & \xrightarrow{\ \pi_*\ } & U \times \mathbf{E} \\
\downarrow & & \downarrow \\
U \times \mathbf{E} & \xrightarrow{\ \ \pi\ \ } & U
\end{array}
$$

The map π_* on each fiber $\mathbf{E} \times \mathbf{E}$ is constant, and is simply the projection of $\mathbf{E} \times \mathbf{E}$ on the first factor $\mathbf{E}$.

Any vector field on $U \times \mathbf{E}$ has a local representation

$$f: U \times \mathbf{E} \to \mathbf{E} \times \mathbf{E}$$

which has therefore two components, $f = (f_1, f_2)$, each f_i mapping $U \times \mathbf{E}$ into $\mathbf{E}$. The next proposition describes second order differential equations and sprays locally in terms of their local representations.

Proposition 5. *Let U be open in the Banach space $\mathbf{E}$, and let $T(U) = U \times \mathbf{E}$ be the tangent bundle. A C^{p-2}-morphism*

$$f \colon U \times \mathbf{E} \to \mathbf{E} \times \mathbf{E}$$

is the local representation of a second order differential equation on U if and only if

$$f(x, v) = \big(v, f_2(x, v)\big).$$

It is a spray if and only if in addition, for all numbers s, we have

$$f_2(x, sv) = s^2 f_2(x, v).$$

Proof. The proof follows at once from the definitions. We leave the symbols to the reader.

We can now write down a spray, letting the local representation be

$$f(x, v) = (v, 0).$$

It satisfies our conditions. We shall see in the chapter on Riemannian metrics how one constructs sprays somewhat less trivially.

We conclude this section by looking more explicitly at the integral curve of a second order differential equation and spray. Let $\varphi = \varphi(t)$ be an integral curve for the second order differential equation ξ. Following the notation of Proposition 5, and thus working locally, our curve has two components,

$$\varphi(t) = \big(y(t), z(t)\big) \in U \times \mathbf{E}.$$

By definition, if f is the local representation of ξ, we must have

$$\frac{d\varphi}{dt} = \left(\frac{dy}{dt}, \frac{dz}{dt}\right) = f(y, z) = \big(z, f_2(y, z)\big).$$

Consequently, our differential equation can be rewritten in the following manner:

$$\frac{dy}{dt} = z$$

$$\frac{d^2y}{dt^2} = \frac{dz}{dt} = f_2\left(y, \frac{dy}{dt}\right)$$

which is of course familiar. The additional condition of a spray simply means that f_2 is homogeneous of degree 2 in the variable dy/dt. It is therefore quadratic. Cf. the remark in Chapter I, §3.

§4. The exponential map

Let X be a manifold and ξ a fixed spray over X. As mentioned before, we let $\mathfrak{D}$ be the set of vectors $v \in T(X)$ such that 1 lies in the domain of β_v. We know that $\mathfrak{D}$ is open, and the exponential map

$$\exp: \mathfrak{D} \to X$$

is given by

$$\exp(v) = \pi\beta_v(1).$$

It is a C^{p-2}-morphism. As usual, we assume X to be a C^p-manifold with $p \geqq 3$.

If $x \in X$ and 0_x denotes the zero vector in T_x, then from **SPR 4**, taking $s = 0$, we see that $\xi(0_x) = 0$. Hence

$$\exp(0_x) = x.$$

Thus our exponential map coincides with π on the zero cross section, and so induces an isomorphism of the cross section onto X. It will be convenient to denote the zero cross section of a vector bundle E over X by $\zeta_E(X)$ or simply ζX if the reference to E is clear. Here, E is the tangent bundle.

We denote by $\exp_x$ the restriction of $\exp$ to the tangent space T_x. Thus

$$\exp_x: T_x \to X.$$

Theorem 8. *Let X be a manifold and ξ a spray on X. Then*

$$\exp_x: T_x \to X$$

induces a local isomorphism at 0_x, and in fact $(\exp_x)_*$ is the identity at 0_x.

Proof. We prove the second assertion first because the main assertion follows from it by the inverse function theorem. Furthermore, since T_x is a vector space, it suffices to determine the derivative of $\exp_x$ on rays, in other words, to determine the derivative with respect to t of a curve $\exp_x(tv)$. This is given by the definition of a spray, and we find

$$\frac{d}{dt}\pi\beta_{tv} = \beta_{tv}.$$

Evaluating this at $t = 0$ and taking into account that β_w has w as initial condition for any w gives us

$$(\exp_x)_*(0_x) = id.$$

§5. *Existence of tubular neighborhoods*

Let X be a submanifold of a manifold Y. A **tubular neighborhood** of X in Y consists of a vector bundle $\pi: E \to X$ over X, an open neighborhood Z of the zero section $\zeta_E X$ in E, and an isomorphism

$$f: Z \to U$$

of Z onto an open set in Y containing X, which commutes with ζ:

We shall call f the **tubular map** and Z or its image $f(Z)$ the corresponding **tube** (in E or Y respectively). The bottom map j is simply the inclusion. We could obviously assume that it is an embedding and define tubular neighborhoods for embeddings in the same way. We shall say that our tubular neighborhood is **total** if $Z = E$. In this section, we investigate conditions under which such neighborhoods exist. We shall consider the uniqueness problem in the next section.

Theorem 9. *Let Y be of class C^p ($p \geq 3$) and admit partitions of unity. Let X be a closed submanifold. Then there exists a tubular neighborhood of X in Y, of class C^{p-2}.*

Proof. Consider the exact sequence of tangent bundles:

$$0 \to T(X) \to T(Y) \mid X \to N(X) \to 0.$$

We know that this sequence splits, and thus there exists some splitting

$$T(Y) \mid X = T(X) \oplus N(X)$$

where $N(X)$ may be identified with a subbundle of $T(Y) \mid X$. We construct a spray ζ on $T(Y)$ using Theorem 7 and obtain the corresponding exponential map. We shall use its restriction to $N(X)$, denoted by $\exp \mid N$. Thus

$$\exp \mid N : \mathfrak{D} \cap N(X) \to Y.$$

We contend that this map is a local isomorphism. To prove this, we may work locally. Corresponding to the submanifold, we have a product decomposition $U = U_1 \times U_2$, with $X = U_1 \times 0$. If U is open in $\mathbf{E}$, then we may take U_1, U_2 open in $\mathbf{F}_1$, $\mathbf{F}_2$ respectively. Then the injection of $N(X)$ in $T(Y) \mid X$ may be represented locally by an exact sequence

$$0 \to U_1 \times \mathbf{F}_2 \xrightarrow{\;\varphi\;} U_1 \times \mathbf{F}_1 \times \mathbf{F}_2,$$

and the inclusion of $T(Y) \mid X$ in $T(Y)$ is simply the inclusion

$$U_1 \times \mathbf{F}_1 \times \mathbf{F}_2 \to U_1 \times U_2 \times \mathbf{F}_1 \times \mathbf{F}_2.$$

We work at the point $(x_1, 0)$ in $U_1 \times \mathbf{F}_2$. We must compute the derivative of the composite map

$$U_1 \times \mathbf{F}_2 \xrightarrow{\;\varphi\;} U_1 \times U_2 \times \mathbf{F}_1 \times \mathbf{F}_2 \xrightarrow{\;\exp\;} Y$$

at $(x_1, 0)$. We can do this by the formula for the partial derivatives. Since the exponential map coincides with the projection on the zero cross section, its "horizontal" partial derivative is the identity. By Theorem 8 of §4 we known that its "vertical" derivative is also the identity. Let $\psi = (\exp) \circ \bar{\varphi}$ (where $\bar{\varphi}$ is simply φ followed by the inclusion). Then for any vector (w_1, w_2) in $\mathbf{F}_1 \times \mathbf{F}_2$ we get

$$D\psi(x_1, 0) \cdot (w_1, w_2) = (w_1, 0) + \varphi_{x_1}(w_2),$$

where φ_{x_1} is the linear map given by φ on the fiber over x_1. By hypothesis, we know that $\mathbf{F}_1 \times \mathbf{F}_2$ is the direct sum of $\mathbf{F}_1 \times 0$ and of the image of φ_{x_1}. This proves that $D\psi(x_1, 0)$ is a toplinear isomorphism, and in fact proves that **the exponential map restricted to a normal bundle is a local isomorphism** on the zero cross section.

We have thus shown that there exists a vector bundle $E \to X$, an open neighborhood Z of the zero section in E, and a mapping $f \colon Z \to Y$ which, for each x in ζ_E, is a local isomorphism at x. We must show that Z can be shrunk so that f restricts to an isomorphism. To do this we follow Godement (*Théorie des Faisceaux*, Hermann, Paris, p. 150). We can find a locally finite open covering of X by open sets U_i in Y such that, for each i we have inverse isomorphisms

$$f_i \colon Z_i \to U_i \qquad \text{and} \qquad g_i \colon U_i \to Z_i$$

between U_i and open sets Z_i in Z, such that each Z_i contains a point x of X, such that f_i, g_i are the identity on X (viewed as a subset of both Z and Y) and such that f_i is the restriction of f to Z_i. We now find a locally finite covering $\{V_i\}$ of X by open sets of Y such that $\bar{V}_i \subset U_i$, and let $V = \bigcup V_i$. We let W be the subset of elements $y \in V$ such that, if y lies in an intersection $\bar{V}_i \cap \bar{V}_j$, then $g_i(y) = g_j(y)$. Then W certainly contains X. We contend that W contains an open subset containing X.

Let $x \in X$. There exists an open neighborhood G_x of x in Y which meets only a finite number of $\bar{V}_i$, say $\bar{V}_{i_1}, \ldots, \bar{V}_{i_r}$. Taking G_x small enough, we can assume that x lies in each one of these, and that G_x is contained in each one of the sets $\bar{U}_{i_1}, \ldots, \bar{U}_{i_r}$. Since x lies in each $\bar{V}_{i_1}, \ldots, \bar{V}_{i_r}$, it is contained in $U_{i_1}, \ldots, U_{i_r}$ and our maps $g_{i_1}, \ldots, g_{i_r}$ take the same value at x, namely x itself. Using the fact that $f_{i_1}, \ldots, f_{i_r}$ are restrictions of f, we see at once that our finite number of maps $g_{i_1}, \ldots, g_{i_r}$ must agree on G_x if we take G_x small enough.

Let G be the union of the G_x. Then G is open, and we can define a map

$$g : G \to g(G) \subset Z$$

by taking g equal to g_i on $G \cap V_i$. Then $g(G)$ is open in Z, and the restriction of f to $g(G)$ is an inverse for g. This proves that f, g are inverse isomorphisms on G and $g(G)$, and concludes the proof of the theorem.

A vector bundle $E \to X$ will be said to be **compressible** if, given an open neighborhood Z of the zero section, there exists an isomorphism

$$\varphi : E \to Z_1$$

of E with an open subset Z_1 of Z containing the zero section, which commutes with the projection on X:

$$E \xrightarrow{\varphi} Z_1$$
$$\searrow \quad \swarrow$$
$$X$$

It is clear that if a bundle is compressible, and if we have a tubular neighborhood defined on Z, then we can get a total tubular neighborhood defined on E. We shall see in the chapter on Riemannian metrics that certain types of vector bundles are compressible (Hilbert bundles, assuming that the base manifold admits partitions of unity).

§6. Uniqueness of tubular neighborhoods

Let X, Y be two manifolds, and $F : \mathbf{R} \times X \to Y$ a morphism. We shall say that F is an **isotopy** (of embeddings) if it satisfies the following conditions. First, for each $t \in \mathbf{R}$, the map F_t given by $F_t(x) = F(t, x)$ is an embedding. Second, there exist numbers $t_0 < t_1$ such that $F_t = F_{t_0}$ for all $t \leq t_0$ and $F_{t_1} = F_t$ for all $t \geq t_1$. We then say that the interval $[t_0, t_1]$ is a **proper** domain for the isotopy, and the constant embeddings on the left and right will also be denoted by $F_{-\infty}$ and $F_{+\infty}$ respectively. We say that two embeddings $f\colon X \to Y$ and $g\colon X \to Y$ are **isotopic** if there exists an isotopy F_t as above such that $f = F_{t_0}$ and $g = F_{t_1}$ (notation as above). We write $f \approx g$ for f isotopic to g.

Using translations of intervals, and multiplication by scalars, we can always transform an isotopy to a new one whose proper domain is contained in the interval $(0, 1)$. Furthermore, the relation of isotopy between embeddings is an equivalence relation. It is obviously symmetric and reflective, and for transitivity, suppose $f \approx g$ and $g \approx h$. We can choose the ranges of these isotopies so that the first one ends and stays constant at g before the second starts moving. Thus it is clear how to compose isotopies in this case.

If $s_0 < s_1$ are two numbers, and $\sigma\colon \mathbf{R} \to \mathbf{R}$ is a function (morphism) such that $\sigma(s) = t_0$ for $s \leq s_0$ and $\sigma(s) = t_1$ for $s \geq s_1$, and σ is monotone increasing, then from a given isotopy F_t we obtain another one, $G_t = F_{\sigma(t)}$. Such a function σ can be used to smooth out a piece of isotopy given only on a closed interval.

Remark. We shall frequently use the following trivial fact: If $f_t\colon X \to Y$ is an isotopy, and if $g\colon X_1 \to X$ and $h\colon Y \to Y_1$ are two embeddings, then the composite map

$$hf_tg\colon X_1 \to Y_1$$

is also an isotopy.

Let Y be a manifold and X a submanifold. Let $\pi\colon E \to X$ be a vector bundle, and Z an open neighborhood of the zero section. An isotopy $f_t\colon Z \to Y$ of open embeddings such that each f_t is a tubular neighborhood of X will be called an **isotopy of tubular neighborhoods**. In what follows, the domain will usually be all of E.

Proposition 6. *Let X be a manifold. Let $\pi\colon E \to X$ and $\pi_1\colon E_1 \to X$ be two vector bundles over X. Let*

$$f\colon E \to E_1$$

be a tubular neighborhood of X in E_1 (identifying X with its zero section in E_1). Then there exists an isotopy

$$f_t\colon E \to E_1$$

with proper domain $[0, 1]$ such that $f_1 = f$ and f_0 is a VB-isomorphism. (If f, π, π_1 are of class C^p then f_t can be chosen of class C^{p-1}.)

Proof. We define F by the formula

$$F_t(e) = t^{-1}f(te)$$

for $t \neq 0$ and $e \in E$. Then F_t is an embedding since it is composed of embeddings (the scalar multiplications by t, t^{-1} are in fact VB-isomorphism).

We must investigate what happens at $t = 0$.

Given $e \in E$, we find an open neighborhood U_1 of πe over which E_1 admits a trivialisation $U_1 \times \mathbf{E}_1$. We then find a still smaller open neighborhood U of πe and an open ball B around 0 in the typical fiber $\mathbf{E}$ of E such that E admits a trivialisation $U \times \mathbf{E}$ over U, and such that the representation $\bar{f}$ of f on $U \times B$ (contained in $U \times \mathbf{E}$) maps $U \times B$ into $U_1 \times \mathbf{E}_1$. This is possible by continuity. On $U \times B$ we can represent $\bar{f}$ by two morphisms,

$$\bar{f}(x, v) = \big(\varphi(x, v), \psi(x, v)\big)$$

and $\varphi(x, 0) = x$ while $\psi(x, 0) = 0$. Observe that for all t sufficiently small, te is contained in $U \times B$ (in the local representation).

We can represent F_t locally on $U \times B$ as the mapping

$$\bar{F}_t(x, v) = \big(\varphi(x, tv), t^{-1}\psi(x, tv)\big).$$

The map φ is then a morphism in the three variables x, v, and t even at $t = 0$. The second component of $\bar{F}_t$ can be written

$$t^{-1}\psi(x, tv) = t^{-1} \int_0^1 D_2\psi(x, stv) \cdot (tv)\, ds$$

and thus t^{-1} cancels t to yield simply

$$\int_0^1 D_2\psi(x, stv) \cdot v\, ds.$$

This is a morphism in t, even at $t = 0$. Furthermore, for $t = 0$, we obtain

$$\bar{F}_0(x, v) = \big(x, D_2\psi(x, 0)v\big).$$

Since f was originally assumed to be an embedding, it follows that $D_2\psi(x, 0)$ is a toplinear isomorphism, and therefore F_0 is a VB-isomorphism. To get our isotopy in standard form, we can use a function $\sigma \colon \mathbf{R} \to \mathbf{R}$ such that $\sigma(t) = 0$ for $t \leq 0$ and $\sigma(t) = 1$ for $t \geq 1$, and σ is monotone increasing. This proves our proposition.

Theorem 10. *Let X be a submanifold of Y. Let $\pi \colon E \to X$ and $\pi_1 \colon E_1 \to X$ be two vector bundles, and assume that E is compressible. Let $f \colon E \to Y$ and $g \colon E_1 \to Y$ be two tubular neighborhoods of X in Y. Then there exists a C^{p-1}-isotopy*

$$f_t \colon E \to Y$$

of tubular neighborhoods with proper domain [0, 1] *and a* VB-*isomorphism* $\lambda \colon E \to E_1$ *such that* $f_1 = f$ *and* $f_0 = g\lambda$.

Proof. We observe that $f(E)$ and $g(E_1)$ are open neighborhoods of X in Y. Let $U = f^{-1}(f(E) \cap g(E_1))$ and let $\varphi \colon E \to U$ be a compression. Let ψ be the composite map

$$E \xrightarrow{\ \varphi\ } U \xrightarrow{\ f \mid U\ } Y$$

$\psi = (f \mid U) \circ \varphi$. Then ψ is a tubular neighborhood, and $\psi(E)$ is contained in $g(E_1)$. Therefore $g^{-1}\psi \colon E \to E_1$ is a tubular neighborhood of the same type considered in the previous proposition. There exists an isotopy of tubular neighborhoods of X:

$$G_t \colon E \to E_1$$

such that $G_1 = g^{-1}\psi$ and G_0 is a VB-isomorphism. Considering the isotopy gG_t, we find an isotopy of tubular neighborhoods

$$\psi_t \colon E \to Y$$

such that $\psi_1 = \psi$ and $\psi_0 = g\omega$ where $\omega \colon E \to E_1$ is a VB-isomorphism. We have thus shown that ψ and $g\omega$ are isotopic (by an isotopy of tubular neighborhoods). Similarly, we see that ψ and $f\mu$ are isotopic for some VB-isomorphism

$$\mu \colon E \to E.$$

Consequently, adjusting the proper domains of our isotopies suitably, we get an isotopy of tubular neighborhoods going from $g\omega$ to $f\mu$, say F_t. Then $F_t\mu^{-1}$ will give us the desired isotopy from $g\omega\mu^{-1}$ to f, and we can put $\lambda = \omega\mu^{-1}$ to conclude the proof.

(By the way, the uniqueness proof did not use the existence theorem for differential equations.)

CHAPTER V

Operations on Vector Fields
and Differential Forms

If $E \to X$ is a vector bundle, then it is of considerable interest to investigate the special operation derived from the functor "multilinear alternating forms." Applying it to the tangent bundle, we call the sections of our new bundle differential forms. One can define formally certain relations between functions, vector fields, and differential forms which lie at the foundations of differential and Riemannian geometry. We shall give the basic system surrounding such forms. In order to have at least one application, we discuss the fundamental 2-form, and in the next chapter connect it with Riemannian metrics in order to construct canonically the spray associated with such a metric.

We assume throughout that our manifolds are Hausdorff, and sufficiently differentiable so that all of our statements make sense.

§1. Vector fields, differential operators, brackets

Let X be a manifold of class C^p and φ a function defined on an open set U, that is a morphism

$$\varphi : U \to \mathbf{R}.$$

Let ξ be a vector field of class C^{p-1}. Recall that

$$T_x\varphi : T_x(U) \to T_x(\mathbf{R}) = \mathbf{R}$$

is a continuous linear map. With it, we shall define a new function to be denoted by $\xi\varphi$ or $\xi(\varphi)$. (There will be no confusion with this notation and composition of mappings.)

Proposition 1. *There exists a unique function $\xi\varphi$ on U of class C^{p-1} such that*

$$(\xi\varphi)(x) = (T_x\varphi)\xi(x).$$

If U is open in the Banach space $\mathbf{E}$ and ξ denotes the local representation of the vector field on U, then

$$(\xi\varphi)(x) = \varphi'(x)\xi(x).$$

Proof. The first formula certainly defines a mapping of U into $\mathbf{R}$. The local formula defines a C^{p-1}-morphism on U. It follows at once from the definitions that the first formula expresses invariantly in terms of the tangent bundle the same mapping as the second. Thus it allows us to define $\xi\varphi$ as a morphism globally, as desired.

Let $\mathfrak{F}^p$ denote the ring of functions (of class C^p). Then our operation $\varphi \mapsto \xi\varphi$ gives rise to a linear map

$$\delta_\xi : \mathfrak{F}^p(U) \to \mathfrak{F}^{p-1}(U),$$

if we let $\delta_\xi\varphi = \xi\varphi$.
A mapping

$$\delta : R \to S$$

from a ring R into an R-algebra S is called a **derivation** if it satisfies the usual formalism: Linearity, and $\delta(ab) = a\delta(b) + \delta(a)b$.

Proposition 2. *Let X be a manifold and U open in X. Let ξ be a vector field over X. If $\delta_\xi = 0$, then $\xi(x) = 0$ for all $x \in U$. Each δ_ξ is a derivation of $\mathfrak{F}^p(U)$ into $\mathfrak{F}^{p-1}(U)$.*

Proof. Suppose $\xi(x) \neq 0$ for some x. We work with the local representations, and take φ to be a continuous linear map of $\mathbf{E}$ into $\mathbf{R}$ such that $\varphi(\xi(x)) \neq 0$, by Hahn-Banach. Then $\varphi'(y) = \varphi$ for all $y \in U$, and we see that $\varphi'(x)\xi(x) \neq 0$, thus proving the first assertion. The second is obvious from the local formula.

From Proposition 2 we deduce that if two vector fields induce the same differential operator on the functions, then they are equal.

Given two vector fields ξ, η on X, we shall now define a new vector field $[\xi, \eta]$.

Proposition 3. *Let ξ, η be two vector fields of class C^{p-1} on X. Then there exists a unique vector field $[\xi, \eta]$ of class C^{p-2} such that for each open set U and function φ on U we have*

$$[\xi, \eta]\varphi = \xi(\eta(\varphi)) - \eta(\xi(\varphi)).$$

If U is open in $\mathbf{E}$ and ξ, η are the local representations of the vector fields, then $[\xi, \eta]$ is given by the local formula

$$[\xi, \eta]\varphi(x) = \varphi'(x)\big(\eta'(x)\xi(x) - \xi'(x)\eta(x)\big).$$

Thus the local representation of $[\xi, \eta]$ is given by

$$[\xi, \eta](x) = \eta'(x)\xi(x) - \xi'(x)\eta(x).$$

Proof. By Proposition 2, any vector field having the desired effect on functions is uniquely determined. We check that the local formula gives us this effect locally. Differentiating formally, we have (using the law for the derivative of a product):

$$(\eta\varphi)'\xi - (\xi\varphi)'\eta = (\varphi'\eta)'\xi - (\varphi'\xi)\eta$$
$$= \varphi'\eta'\xi + \varphi''\eta\xi - \varphi'\xi'\eta - \varphi''\xi\eta.$$

The terms involving φ'' must be understood correctly. For instance, the first such term at a point x is simply

$$\varphi''(x)\big(\eta(x), \xi(x)\big)$$

remembering that φ'' is a bilinear map, and can thus be evaluated at the two vectors $\eta(x)$ and $\xi(x)$. However, we know that $\varphi''(x)$ is symmetric. Hence the two terms involving the second derivative of φ cancel, and give us our formula.

Corollary. *The bracket $[\xi, \eta]$ is bilinear in both arguments, we have $[\xi, \eta] = -[\eta, \xi]$, and Jacobi's identity*

$$[\xi, [\eta, \zeta]] + [\eta, [\zeta, \xi]] + [\zeta, [\xi, \eta]] = 0.$$

Proof. The first two assertions are obvious. The third comes from the definition of the bracket. We apply the vector field on the left of the equality to a function φ. All the terms cancel out (the reader will write it out as well or better than the author).

We make some comments concerning the functoriality of vector fields. Let

$$f \colon X \to Y$$

be an isomorphism. Let ξ be a vector field over X. Then we obtain an induced vector field $f_*\xi$ over Y, defined by the formula

$$(f_*\xi)\big(f(x)\big) = Tf(\xi(x)).$$

It is the vector field making the following diagram commutative.

$$\begin{array}{ccc} TX & \xrightarrow{Tf} & TY \\ \xi \big\uparrow\big\uparrow & & \big\uparrow\big\uparrow {\scriptstyle f_*\xi} \\ X & \xrightarrow{\;\;f\;\;} & Y \end{array}$$

We shall also write f^* for $(f^{-1})_*$ when applied to a vector field. Thus we have the formulas

$$f_*\xi = Tf \circ \xi \circ f^{-1} \qquad \text{and} \qquad f^*\xi = Tf^{-1} \circ \xi \circ f.$$

If f is not an isomorphism, then one cannot in general define the direct or inverse image of a vector field as done above. However, let ξ be a vector field over X, and let η be a vector field over Y. If for each $x \in X$ we have

$$Tf(\xi(x)) = \eta(f(x)),$$

then we shall say that f maps ξ into η, or that ξ and η are f-related. If this is the case, then we may denote by $f_*\xi$ the map from $f(X)$ into TY defined by the above formula.

Let ξ_1, ξ_2 be vector fields over X, and let η_1, η_2 be vector fields over Y. If ξ_i is f-related to η_i for $i = 1, 2$ then as maps on $f(X)$ we have

$$f_*[\xi_1, \xi_2] = [\eta_1, \eta_2].$$

We may write suggestively the formula in the form

$$f_*[\xi_1, \xi_2] = [f_*\xi_1, f_*\xi_2].$$

Of course, this is meaningless in general, since $f_*\xi_1$ may not be a vector field on Y. When f is an isomorphism, then it is a correct formulation of the other formula. In any case, it suggests the correct formula.

To prove the formula, we work with the local representations, when $X = U$ is open in $\mathbf{E}$, and $Y = V$ is open in $\mathbf{F}$. Then ξ_i, η_i are maps of U, V into the spaces $\mathbf{E}$, $\mathbf{F}$ respectively. For $x \in X$ we have

$$(f_*[\xi_1, \xi_2])(x) = f'(x)\big(\xi_2'(x)\xi_1(x) - \xi_1'(x)\xi_2(x)\big).$$

On the other hand, by assumption, we have

$$\eta_i(f(x)) = f'(x)\xi_i(x),$$

so that

$$
\begin{aligned}
[\eta_1, \eta_2](f(x)) &= \eta_2'(f(x))\eta_1(f(x)) - \eta_1'(f(x))\eta_2(f(x)) \\
&= \eta_2'(f(x))f'(x)\xi_1(x) - \eta_1'(f(x))f'(x)\xi_2(x) \\
&= (\eta_2 \circ f)'(x)\xi_1(x) - (\eta_1 \circ f('(x)\xi_2(x))) \\
&= f''(x) \cdot \xi_2(x) \cdot \xi_1(x) + f'(x)\xi_2'(x)\xi_1(x) \\
&\quad -f''(x) \cdot \xi_1(x) \cdot \xi_2(x) - f'(x)\xi('_1 x)\xi_2(x).
\end{aligned}
$$

Since $f''(x)$ is symmetric, two terms cancel, and the remaining two terms give the same value as $(f_*[\xi_1, \xi_2])(x)$, as was to be shown.

The bracket between vector fields gives an infinitesimal criterion for commutativity in various contexts. We give here one theorem of a general nature as an example of this phenomenon.

Theorem 1. *Let* ξ, η *be vector fields on* X, *and assume that* $[\xi, \eta] = 0$. *Let* α *and* β *be the flows for* ξ *and* η *respectively. Then for real values* t, s *we have*

$$\alpha_t \circ \beta_s = \beta_s \circ \alpha_t.$$

Or in other words, for any $x \in X$ *we have*

$$\alpha(t, \beta(s, x)) = \beta(s, \alpha(t, x)),$$

in the sense that if for some value of t *a value of* s *is in the domain of one of these expressions, then it is in the domain of the other and the two expressions are equal.*

Proof. For a fixed value of t, the two curves in s given by the right- and left-hand side of the last formula have the same initial condition, namely $\alpha_t(x)$. The curve on the right

$$s \mapsto \beta(s, \alpha(t, x))$$

is by definition the integral curve of η. The curve on the left

$$s \mapsto \alpha(t, \beta(s, x))$$

is the image under α_t of the integral curve for η having initial condition x. Since x is fixed, let us denote $\beta(s, x)$ simply by $\beta(s)$. What we must show is that the two curves on the right and on the left satisfy the same differential equation.

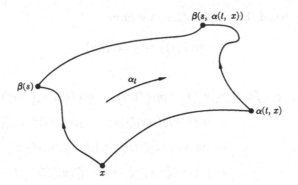

In the above figure, we see that the flow α_t shoves the curve on the left to the curve on the right. We must compute the tangent vectors to the curve on the right. We have

$$\frac{d}{ds}\left(\alpha_t(\beta(s))\right) = D_2\alpha(t, \beta(s))\beta'(s)$$

$$= D_2\alpha(t, \beta(s))\eta(\beta(s)).$$

Now fix s, and denote this last expression by $F(t)$. We must show that if

$$G(t) = \eta(\alpha(t, \beta(s))),$$

then

$$F(t) = G(t).$$

We have trivially $F(0) = G(0)$, in other words the curves F and G have the same initial condition. On the other hand,

$$F'(t) = \xi'(\alpha(t, \beta(s)))D_2\alpha(t, \beta(s))\eta(\beta(s))$$

and

$$G'(t) = \eta'(\alpha(t, \beta(s)))\xi(\alpha(t, \beta(s)))$$

$$= \xi'(\alpha(t, \beta(s)))\eta(\alpha(t, \beta(s)))   \text{(because } [\xi, \eta] = 0).$$

Hence we see that our two curves F and G satisfy the same differential equation, whence they are equal. This proves our theorem.

Vector fields ξ, η such that $[\xi, \eta] = 0$ are said to **commute**. One can generalize the process of straightening out vector fields to a finite number of commuting vector fields, using the same method of proof, using Theorem 1. As another application, one can prove that if the Lie algebra of a connected Lie group is commutative, then the group is commutative. Cf. the section on Lie groups.

§2. *Lie derivative*

Let λ be a differentiable functor on Banach spaces. For convenience, take λ to be covariant and in one variable. What we shall say in the rest of this section would hold in the same way (with slightly more involved notation) if λ had several variables and were covariant in some and contravariant in others.

Given a manifold X, we can take $\lambda(T(X))$. It is a vector bundle over X, which we denote by $T_\lambda(X)$ as in Chapter III. Its sections $\Gamma_\lambda(X)$ are the tensor fields of type λ.

Let ξ be a vector field on X, and U open in X. It is then possible to associate with ξ a map

$$\mathscr{L}_\xi : \Gamma_\lambda(U) \to \Gamma_\lambda(U)$$

(with a loss of two derivatives). This is done as follows.

Given a point x of U and a local flow α for ξ at x, we have for each t sufficiently small a local isomorphism α_t in a neighborhood of our point x. Recall that locally, $\alpha_t^{-1} = \alpha_{-t}$. If η is a tensor field of type λ, then the composite mapping $\eta \circ \alpha_t$ has its range in $T_\lambda(X)$. Finally, we can take the tangent map $T(\alpha_{-t}) = (\alpha_{-t})_*$ to return to $T_\lambda(X)$ in the fiber above x. We thus obtain a composite map

$$F(t, x) = (\alpha_{-t})_* \circ \eta \circ \alpha_t(x) = (\alpha_t^* \eta)(x),$$

which is a morphism, locally at x. We take its derivative with respect to t and evaluate it at 0. After looking at the situation locally in a trivialisation of $T(X)$ and $T_\lambda(X)$ at x, one sees that the map one obtains gives a section of $T_\lambda(U)$, that is a tensor field of type λ over U. This is our map $\mathscr{L}_\xi$. To summarize,

$$\mathscr{L}_\xi \eta = \frac{d}{dt}\bigg|_{t=0} (\alpha_{-t})_* \circ \eta \circ \alpha_t.$$

It is then an exercise to show that the operation of vector fields on functions, and the bracket of vector fields are special cases of the general operation θ_λ. In the first case, one takes for λ the constant functor $\mathbf{R}$, and in the second case, the identity.

This map $\mathscr{L}_\xi$ is called the **Lie derivative**. In each case we shall carry out the exercise, beginning now with functions and vector fields, and continuing later with differential forms.

First let φ be a function. Then by the general definition, the Lie derivative of this function with respect to the vector field ξ with flow α is defined to be

$$\mathscr{L}_\xi \varphi(x) = \lim_{t \to 0} \frac{1}{t} [\varphi(\alpha(t, x)) - \varphi(x)],$$

or in other words,

$$\mathcal{L}_\xi \varphi = \frac{d}{dt} \left(\alpha_t^* \varphi \right) \bigg|_{t=0}.$$

Our assertion is then that

$$\boxed{\mathcal{L}_\xi \varphi = \xi \varphi.}$$

To prove this, let

$$F(t) = \varphi \big(\alpha(t, x) \big).$$

Then

$$
\begin{aligned}
F'(t) &= \varphi' \big(\alpha(t, x) \big) D_1 \alpha(t, x) \\
&= \varphi' \big(\alpha(t, x) \big) \xi \big(\alpha(t, x) \big),
\end{aligned}
$$

because α is a flow for ξ. Using the initial condition at $t = 0$, we find that

$$F'(0) = \varphi'(x) \xi(x),$$

which is precisely the value of $\xi \varphi$ at x, thus proving our assertion.

If ξ, η are vector fields, then

$$\boxed{\mathcal{L}_\xi \eta = [\xi, \eta].}$$

As before, let α be a flow for ξ. The Lie derivative is given by

$$\mathcal{L}_\xi \eta = \frac{d}{dt} \left(\alpha_t^* \eta \right) \bigg|_{t=0}.$$

Letting ξ and η denote the local representations of the vector fields, we note that the local representation of $(\alpha_t^* \eta)(x)$ is given by

$$(\alpha_t^* \eta)(x) = F(t) = D_2 \alpha(-t, x) \eta \big(\alpha(t, x) \big).$$

We must therefore compute $F'(t)$, and then $F'(0)$. Using the chain rule, the formula for the derivative of a product, and the differential equation satisfied by $D_2 \alpha$, we obtain

$$
\begin{aligned}
F'(t) &= -D_1 D_2 \alpha(-t, x) \eta \big(\alpha(t, x) \big) + D_2 \alpha(-t, x) \eta' \big(\alpha(t, x) \big) D_1 \alpha(t, x) \\
&= -\xi' \big(\alpha(-t, x) \big) D_2 \alpha(-t, x) \eta \big(\alpha(t, x) \big) + D_2 \alpha(-t, x) \eta' \big(\alpha(t, x) \big).
\end{aligned}
$$

Putting $t = 0$ proves our formula, taking into account the initial conditions

$$\alpha(0, x) = x \quad \text{and} \quad D_2\alpha(0, x) = id.$$

§3. Exterior derivative

Let X be a manifold. The functor $L_a^r(\mathbf{E})$ (r-multilinear continuous alternating forms) extends to arbitrary vector bundles, and in particular, to the tangent bundle of X. A **differential form** of degree r, or simply an **r-form** on X, is a cross section of $L_a^r(T(X))$, that is a tensor field of type L_a^r. If X is of class C^p, forms will be assumed to be of a suitable class C^s with $1 \leqq s \leqq p - 1$. They are denoted by $\Omega^r(X)$. If ω is an r-form, then $\omega(x)$ is an element of $L_a^r(T_x(X))$, and is thus an r-multilinear alternating form of $T_x(X)$ into $\mathbf{R}$. We sometimes denote $\omega(x)$ by ω_x.

Suppose U is open in the Banach space $\mathbf{E}$. Then $L_a^r(T(U))$ is equal to $U \times L_a^r(\mathbf{E})$ and a differential form is entirely described by the projection on the second factor, which we call its local representation, following our general system (Chapter III, §4). Such a local representation is therefore a morphism

$$\omega: U \to L_a^r(\mathbf{E}).$$

Let ω be in $L_a^r(\mathbf{E})$ and $v_1, \ldots, v_r$ elements of $\mathbf{E}$. We denote the value $\omega(v_1, \ldots, v_r)$ also by

$$\langle \omega, v_1 \times \cdots \times v_r \rangle.$$

Similarly, let $\xi_1, \ldots, \xi_r$ be vector fields on an open set U, and let ω be an r-form on X. We denote by

$$\langle \omega, \xi_1 \times \cdots \times \xi_r \rangle$$

the mapping from U into $\mathbf{R}$ whose value at a point x in U is

$$\langle \omega(x), \xi_1(x) \times \cdots \times \xi_r(x) \rangle.$$

Looking at the situation locally on an open set U such that $T(U)$ is trivial, we see at once that this mapping is a morphism (i.e. a function on U) of the same degree of differentiability as ω and the ξ_i.

Proposition 4. *Let x_0 be a point of X and ω an r-form on X. If*

$$\langle \omega, \xi_1 \times \cdots \times \xi_r \rangle(x_0)$$

is equal to 0 for all vector fields $\xi_1, \ldots, \xi_r$ at x_0 (i.e. defined on some neighborhood of x_0), then $\omega(x_0) = 0$.

Proof. Considering things locally in terms of their local representations, we see that if $\omega(x_0)$ is not 0, then it does not vanish at some r-tuple of vectors $(v_1, \ldots, v_r)$. We can take vector fields at x_0 which take on these values at x_0 and from this our assertion is obvious.

It is convenient to agree that a differential form of degree 0 is a function. In the next proposition, we describe the exterior derivative of an r-form, and it is convenient to describe this situation separately in the case of functions.

Therefore let $f\colon X \to \mathbf{R}$ be a function. For each $x \in X$, the tangent map

$$T_x f\colon T_x(X) \to T_{f(x)}(\mathbf{R}) = \mathbf{R}$$

is a continuous linear map, and looking at local representations shows at once that the collection of such maps defines a 1-form which will be denoted by df. Furthermore, from the definition of the operations of vector fields on functions, it is clear that df is the unique 1-form such that for every vector field ξ we have

$$\langle df, \xi \rangle = \xi f.$$

To extend the definition of d to forms of higher degree, we recall that if

$$\omega\colon U \to L_a^r(\mathbf{E})$$

is the local representation of an r-form over an open set U of $\mathbf{E}$, then for each x in U,

$$\omega'(x)\colon \mathbf{E} \to L_a^r(\mathbf{E})$$

is a continuous linear map. Applied to a vector v in $\mathbf{E}$, it therefore gives rise to an r-form on $\mathbf{E}$.

Proposition 5. *Let ω be an r-form of class C^{p-1} on X. Then there exists a unique $(r+1)$-form $d\omega$ on X of class C^{p-2} such that, for any open set U of X and vector fields $\xi_0, \ldots, \xi_r$ on U we have*

$$\langle d\omega, \xi_0 \times \cdots \times \xi_r \rangle = \sum_{i=0}^{r} (-1)^i \xi_i \langle \omega, \xi_0 \times \cdots \times \hat{\xi}_i \times \cdots \times \xi_r \rangle$$

$$+ \sum_{i<j} (-1)^{i+j} \langle \omega, [\xi_i, \xi_j] \times \xi_0 \times \cdots \times \hat{\xi}_i \times \cdots \times \hat{\xi}_j \times \cdots \times \xi_r \rangle.$$

If furthermore U is open in $\mathbf{E}$ and $\omega, \xi_0, \ldots, \xi_r$ are the local representations of the form and the vector fields respectively, then at a point x the value of the expression above is equal to

$$\sum_{i=0}^{r} (-1)^i \langle \omega'(x)\xi_i(x), \xi_0(x) \times \cdots \times \widehat{\xi_i(x)} \times \cdots \times \xi_r(x) \rangle.$$

Proof. As before, we observe that the local formula defines a differential form. If we can prove that it gives the same thing as the first formula, which is expressed invariantly, then we can globalize it, and we are done. Let us denote by S_1 and S_2 the two sums occurring in the invariant expression, and let L be the local expression. We must show that $S_1 + S_2 = L$. We consider S_1, and apply the definition of ξ_i operating on a function locally, as in Proposition 1, at a point x. We obtain

$$S_1 = \sum_{i=0}^{r} (-1)^i \langle \omega, \xi_0 \times \cdots \times \hat{\xi}_i \times \cdots \times \xi_r \rangle'(x) \xi_i(x).$$

The derivative is perhaps best computed by going back to the definition. Applying this definition directly, and discarding second order terms, we find that S_1 is equal to

$$\sum (-1)^i \langle \omega'(x)\xi_i(x), \xi_0(x) \times \cdots \times \widehat{\xi_i(x)} \times \cdots \times \xi_r(r) \rangle$$
$$+ \sum_i \sum_{j<i} \langle \omega(x), \xi_0(x) \times \cdots \times \xi_j'(x)\xi_i(x) \times \cdots \times \widehat{\xi_i(x)} \times \cdots \times \xi_r(x) \rangle$$
$$+ \sum_i \sum_{j<i} \langle \omega(x), \xi_0(x) \times \cdots \times \widehat{\xi_i(x)} \times \cdots \times \xi_j'(x)\xi_i(x) \times \cdots \times \xi_r(x) \rangle.$$

Of these three sums, the first one is the local formula L. As for the other two, permuting j and i in the last, and moving the term $\xi_j'(x)\xi_i(x)$ to the first position, we see that they combine to give (symbolically)

$$-\sum_i \sum_{j<i} (-1)^{i+j} \langle \omega, (\xi_j'\xi_i - \xi_j'\xi_j) \times \xi_0 \times \cdots \times \hat{\xi}_i \times \cdots \times \hat{\xi}_j \times \cdots \times \xi_r \rangle$$

(evaluated at x). Using Proposition 3, we see that this combination is equal to $-S_2$. This proves that $S_1 + S_2 = L$, as desired.

We call $d\omega$ the **exterior derivative** of ω.

Let ω, ψ be continuous multilinear alternating forms of degree r and s respectively on the Banach space **E**. In multilinear algebra, one defines their **wedge product** as an $(r + s)$-continuous multilinear alternating form, by the formula

$$(\omega \wedge \psi)(v_1, \ldots, v_{r+s}) \frac{1}{(r + s)!} = \sum \varepsilon(\sigma) \omega(v_{\sigma 1}, \ldots, v_{\sigma r}) \psi(v_{\sigma(r+1)}, \ldots, v_{\sigma s})$$

the sum being taken over all permutations σ of $(1, \ldots, r + s)$. This definition extends at once to differential forms on a manifold, if we view it as giving the value for $\omega \wedge \psi$ at a point x. The v_i are then elements of the tangent space T_x, and considering the local representations shows at once that the wedge product so defined gives a morphism of the manifold X into $L_a^{r+s}(T(X))$, and is therefore a differential form.

If we regard functions on X as differential forms of degree 0, then the ordinary product of a function by a differential form can be viewed as the wedge product. Thus if f is a function and ω a differential form, then

$$f\omega = f \wedge \omega.$$

(The form on the left has the value $f(x)\omega(x)$ at x.)

The next proposition gives us more formulas concerning differential forms.

Proposition 6. *Let* ω, ψ *be differential forms on* X. *Then*

EXD 1. $d(\omega \wedge \psi) = d\omega \wedge \psi + (-1)^{\deg(\omega)}\omega \wedge d\psi$.

EXD 2. $dd\omega = 0$ *(with enough differentiability, say $p \geqq 4$).*

Proof. This is a simple formal exercise in the use of the local formula for the local representation of the exterior derivative. We leave it to the reader.

When the manifold is finite dimensional, then one can give a local representation for differential forms and the exterior derivative in terms of local coordinates, which are especially useful in integration which fits the notation better. We shall therefore carry out this local formulation in full.

We recall first two simple results from linear (or rather multilinear) algebra. We use the notation $\mathbf{E}^{(r)} = \mathbf{E} \times \mathbf{E} \times \cdots \times \mathbf{E}$, r times.

Theorem A. *Let* $\mathbf{E}$ *be a finite dimensional vector space over the reals of dimension n. For each positive integer r with $1 \leqq r \leqq n$ there exists a vector space $\wedge^r\mathbf{E}$ and a multilinear alternating map*

$$\mathbf{E}^{(r)} \to \wedge^r\mathbf{E}$$

denoted by $(u_1, \ldots, u_r) \mapsto u_1 \wedge \cdots \wedge u_r$, *having the following property: If* $\{v_1, \ldots, v_n\}$ *is a basis of* $\mathbf{E}$, *then the elements*

$$\{v_{i_1} \wedge \cdots \wedge v_{i_r}\}, \qquad i_1 < i_2 < \cdots < i_r,$$

form a basis of $\wedge^r\mathbf{E}$.

We recall that **alternating** means that $u_1 \wedge \cdots \wedge u_r = 0$ if $u_i = u_j$ for some $i \neq j$. We call $\wedge^r\mathbf{E}$ the rth **alternating** product (or **exterior** product) on $\mathbf{E}$. If $r = 0$, we define $\wedge^0\mathbf{E} = \mathbf{R}$. Elements of $\wedge^r\mathbf{E}$ which can be written in the form $u_1 \wedge \cdots \wedge u_r$ are called **decomposable**. Such elements generate $\wedge^r\mathbf{E}$. If $r > \dim E$, we define $\wedge^r\mathbf{E} = \{0\}$.

Theorem B. *For each pair of positive integers (r, s), there exists a unique product (bilinear map)*

$$\wedge^r E \times \wedge^s E \to \wedge^{r+s} E$$

such that if $u_1, \ldots, u_r, w_1, \ldots, w_s \in E$ then

$$(u_1 \wedge \cdots \wedge u_r) \times (w_1 \wedge \cdots \wedge w_s) \mapsto u_1 \wedge \cdots \wedge u_r \wedge w_1 \wedge \cdots \wedge w_s.$$

This product is associative.

The proofs for these two statements can be found, for instance, in my *Linear Algebra*.

Let E^* be the dual space, $E^* = L(E, \mathbf{R})$. If $E = \mathbf{R}^n$ and $\lambda_1, \ldots, \lambda_n$ are the coordinate functions, then each λ_i is an element of the dual space, and in fact $\{\lambda_1, \ldots, \lambda_n\}$ is a basis of this dual space. Let $E = \mathbf{R}^n$. There is an isomorphism

$$\boxed{\wedge^r E^* \xrightarrow{\approx} L_a^r(E, \mathbf{R})}$$

given in the following manner. If $g_1, \ldots, g_r \in E^*$ and $v_1, \ldots, v_r \in E$, then the value

$$\det \big(g_i(v_j)\big)$$

is multilinear alternating both as a function of $(g_1, \ldots, g_r)$ and $(v_1, \ldots, v_r)$. Thus it induces a pairing

$$\wedge^r E^* \times E^r \to \mathbf{R}$$

and a map

$$\wedge^r E^* \to L_a^r(E, \mathbf{R}).$$

This map is the isomorphism mentioned above. Using bases, it is easy to verify that it is an isomorphism (at the level of elementary algebra).

Thus in the finite dimensional case, we may identify $L_a^r(E, \mathbf{R})$ with the alternating product $\wedge^r E^*$, and consequently we may view the local representation of a differential form of degree r to be a map

$$\omega: U \to \wedge^r E^*$$

from U into the rth alternating product of E^*. We say that the form is of class C^p if the map is of class C^p. (We view $\wedge^r E^*$ as a normed vector space, using any norm. It does not matter which, since all norms on a finite dimensional vector space are equivalent.)

Since $\{\lambda_1, \ldots, \lambda_n\}$ is a basis of $\mathbf{E}^*$, we can express each differential form in terms of its coordinate functions with respect to the basis

$$\{\lambda_{i_1} \wedge \cdots \wedge \lambda_{i_r}\}, \qquad\qquad (i_1 < \cdots < i_r),$$

namely for each $x \in U$ we have

$$\omega(x) = \sum_{(i)} f_{i_1 \cdots i_r}(x) \lambda_{i_1} \wedge \cdots \wedge \lambda_{i_r},$$

where $f_{(i)} = f_{i_1 \cdots i_r}$ is a function on U. Each such function has the same order of differentiability as ω. We call the preceding expression the **standard form** of ω. We say that a form is **decomposable** if it can be written as just one term $f(x)\lambda_{i_1} \wedge \cdots \wedge \lambda_{i_r}$. Every differential form is a sum of decomposable ones.

We agree to the convention that functions are differential forms of degree 0.

It is clear that the differential forms of given degree r form a vector space, denoted by $\Omega^r(U)$.

Let $\mathbf{E} = \mathbf{R}^n$. Let f be a function on U. For each $x \in U$ the derivative

$$f'(x) \colon \mathbf{R}^n \to \mathbf{R}$$

is a linear map, and thus an element of the dual space. Thus

$$f' \colon U \to \mathbf{E}^*$$

represents a differential form of degree 1, which is usually denoted by df. If f is of class C^p, then df is class C^{p-1}.

Let λ_i be the ith coordinate function. Then we know that

$$d\lambda_i(x) = \lambda_i'(x) = \lambda_i$$

for each $x \in U$ because $\lambda'(x) = \lambda$ for any continuous linear map λ. Whenever $\{x_1, \ldots, x_n\}$ are used systematically for the coordinates of a point in $\mathbf{R}^n$, it is customary in the literature to use the notation

$$d\lambda_i(x) = dx_i.$$

This is slightly incorrect, but is useful in formal computations. We shall also use it in this book on occasions. Similarly, we also write (incorrectly)

$$\omega = \sum_{(i)} f_{(i)} \, dx_{i_1} \wedge \cdots \wedge dx_{i_r}$$

instead of the correct

$$\omega(x) = \sum_{(i)} f_{(i)}(x) \lambda_{i_1} \wedge \cdots \wedge \lambda_{i_r}.$$

In terms of coordinates, the map df (or f') is given by

$$df(x) = f'(x) = D_1 f(x)\lambda_1 + \cdots + D_n f(x)\lambda_n$$

where $D_i f(x) = \partial f/\partial x_i$ is the ith partial derivative. This is simply a restatement of the fact that if $h = (h_1, \ldots, h_n)$ is a vector, then

$$f'(x)h = \frac{\partial f}{\partial x_1} h_1 + \cdots + \frac{\partial f}{\partial x_n} h_n.$$

Thus in old notation, we have

$$df(x) = \frac{\partial f}{\partial x_1} dx_1 + \cdots + \frac{\partial f}{\partial x_n} dx_n.$$

We shall develop the theory of the alternating product and the exterior derivative directly without assuming Propositions 5 or 6 in the finite dimensional case.

Let ω and ψ be forms of degrees r and s respectively, on the open set U. For each $x \in U$ we can then take the alternating product $\omega(x) \wedge \psi(x)$ and we define the **alternating product** $\omega \wedge \psi$ by

$$(\omega \wedge \psi)(x) = \omega(x) \wedge \psi(x).$$

(It is an exercise to verify that this product corresponds to the product defined previously before Proposition 6 under the isomorphism between $L_a^r(\mathbf{E}, \mathbf{R})$ and the rth alternating product in the finite dimensional case.) If f is a differential form of degree 0, that is a function, then we have again

$$f \wedge \omega = f\omega,$$

where $(f\omega)(x) = f(x)\omega(x)$. By definition, we then have

$$\omega \wedge f\psi = f\omega \wedge \psi.$$

We shall now define the **exterior derivative** $d\omega$ for any differential form ω. We have already done it for functions. We shall do it in general first in terms of coordinates, and then show that there is a characterization independent of these coordinates. If

$$\omega = \sum_{(i)} f_{(i)} \, d\lambda_{i_1} \wedge \cdots \wedge d\lambda_{i_r},$$

we define

$$d\omega = \sum_{(i)} df_{(i)} \wedge d\lambda_{i_1} \wedge \cdots \wedge d\lambda_{i_r}.$$

Example. Suppose $n = 2$ and ω is a 1-form, given in terms of the two coordinates (x, y) by

$$\omega(x, y) = f(x, y)\, dx + g(x, y)\, dy.$$

Then

$$d\omega(x, y) = df(x, y) \wedge dx + dg(x, y) \wedge dy$$

$$= \left(\frac{\partial f}{\partial x}\, dx + \frac{\partial f}{\partial y}\, dy\right) \wedge dx + \left(\frac{\partial g}{\partial x}\, dx + \frac{\partial g}{\partial y}\, dy\right) \wedge dy$$

$$= \frac{\partial f}{\partial y}\, dy \wedge dx + \frac{\partial g}{\partial x}\, dx \wedge dy$$

$$= \left(\frac{\partial f}{\partial y} - \frac{\partial g}{\partial x}\right) dy \wedge dx$$

because the terms involving $dx \wedge dx$ and $dy \wedge dy$ are equal to 0.

The map d is linear, and satisfies

$$d(\omega \wedge \psi) = d\omega \wedge \psi + (-1)^r \omega \wedge d\psi$$

if $r = \deg \omega$. The map d is uniquely determined by these properties, and by the fact that for a function f, we have $df = f'$.

Proof. The linearity of d is obvious. Hence it suffices to prove the formula for decomposable forms. We note that for any function f we have

$$d(f\omega) = df \wedge \omega + f\, d\omega.$$

Indeed, if ω is a function g, then from the derivative of a product we get $d(fg) = f\, dg + g\, df$. If

$$\omega = g\, d\lambda_{i_1} \wedge \cdots \wedge d\lambda_{i_r}$$

where g is a function, then

$$d(f\omega) = d(fg\, d\lambda_{i_1} \wedge \cdots \wedge d\lambda_{i_r}) = d(fg) \wedge d\lambda_{i_1} \wedge \cdots \wedge d\lambda_{i_r}$$

$$= (f\, dg + g\, df) \wedge d\lambda_{i_1} \wedge \cdots \wedge d\lambda_{i_r}$$

$$= f\, d\omega + df \wedge \omega,$$

as desired. Now suppose that

$$\omega = f\, d\lambda_{i_1} \wedge \cdots \wedge d\lambda_{i_r} \qquad \text{and} \qquad \psi = g\, d\lambda_{j_1} \wedge \cdots \wedge d\lambda_{js}$$

$$= f\tilde{\omega} \qquad\qquad\qquad\qquad\qquad = g\tilde{\psi}$$

with $i_1 < \cdots < i_r$ and $j_1 < \cdots < j_s$ as usual. If some $i_\nu = j_\mu$, then from the definitions we see that the expressions on both sides of the equality in the theorem are equal to 0. Hence we may assume that the sets of indices $i_1, \ldots, i_r$ and $j_1, \ldots, j_s$ have no element in common. Then $d(\tilde{\omega} \wedge \tilde{\psi}) = 0$ by definition, and

$$
\begin{aligned}
d(\omega \wedge \psi) = d(fg\tilde{\omega} \wedge \tilde{\psi}) &= d(fg) \wedge \tilde{\omega} \wedge \tilde{\psi} \\
&= (g\, df + f\, dg) \wedge \tilde{\omega} \wedge \tilde{\psi} \\
&= d\omega \wedge \psi + f\, dg \wedge \tilde{\omega} \wedge \tilde{\psi} \\
&= d\omega \wedge \psi + (-1)^r f\tilde{\omega} \wedge dg \wedge \tilde{\psi} \\
&= d\omega \wedge \psi + (-1)^r \omega \wedge d\psi,
\end{aligned}
$$

thus proving the desired formula, in the present case. (We used the fact that $dg \wedge \tilde{\omega} = (-1)^r \tilde{\omega} \wedge dg$ whose proof is left to the reader.) The formula in the general case follows because any differential form can be expressed as a sum of forms of the type just considered, and one can then use the bilinearity of the product. Finally, d is uniquely determined by the formula, and its effect on functions, because any differential form is a sum of forms of type $f\, d\lambda_{i_1} \wedge \cdots \wedge d\lambda_{i_r}$ and the formula gives an expression of d in terms of its effect on forms of lower degree. By induction, if the value of d on functions is known, its value can then be determined on forms of degree ≥ 1. This proves our assertion.

Let ω be a form of class C^2. Then $dd\omega = 0$.

Proof. If f is a function, then

$$
df(x) = \sum_{j=1}^{n} \frac{\partial f}{\partial x_j}\, dx_j
$$

and

$$
ddf(x) = \sum_{j=1}^{n} \sum_{k=1}^{n} \frac{\partial^2 f}{\partial x_k\, \partial x_j}\, dx_k \wedge dx_j.
$$

Using the fact that the partials commute, and the fact that for any two positive integers r, s we have $dx_r \wedge dx_s = -dx_s \wedge dx_r$, we see that the preceding double sum is equal to 0. A similar argument shows that the theorem is true for 1-forms, of type $g(x)\, dx_i$ where g is a function, and thus for all 1-forms by linearity. We proceed by induction. It suffices to prove the formula in general for decomposable forms. Let ω be decomposable of degree r, and write

$$
\omega = \eta \wedge \psi,
$$

where deg $\psi = 1$. Using the formula for the derivative of an alternating product twice, and the fact that $dd\psi = 0$ and $dd\eta = 0$ by induction, we see at once that $dd\omega = 0$, as was to be shown.

We conclude this section by giving some properties of the pull-back of forms. As we saw at the end of Chapter III, §4, if $f: X \to Y$ is a morphism and if ω is a differential form on Y, then we get a differential form $f^*(\omega)$ on X, which is given at a point $x \in X$ by the formula

$$f^*(\omega)_x = \omega_{f(x)} \circ (T_xf)',$$

if ω is of degree r. This holds for $r \geqq 1$. The corresponding local representation formula reads

$$\langle f^*\omega(x), \xi_1(x) \times \cdots \times \xi_r(x) \rangle = \langle \omega(f(x)), f'(x)\xi_1(x) \times \cdots \times f'(x)\xi_r(x) \rangle$$

if $\xi_1, \ldots, \xi_r$ are vector fields.

In the case of a 0-form, that is a function, its pull-back is simply the composite function. In other words, if φ is a function on Y, viewed as a form of degree 0, then

$$f^*(\varphi) = \varphi \circ f.$$

It is clear that the pull-back is linear, and satisfies the following properties.

Property 1. *If ω, ψ are two differential forms on Y, then*

$$f^*(\omega \wedge \psi) = f^*(\omega) \wedge f^*(\psi).$$

Property 2. *If ω is a differential form on Y, then*

$$df^*(\omega) = f^*(d\omega).$$

Property 3. *If $f: X \to Y$ and $g: Y \to Z$ are two morphisms, and ω is a differential form on Z, then*

$$f^*(g^*(\omega)) = (g \circ f)^*(\omega).$$

Finally, in the case of forms of degree 0:

Property 4. *If $f: X \to Y$ is a morphism, and g is a function on Y, then*

$$d(g \circ f) = f^*(dg)$$

and at a point $x \in X$, the value of this 1-form is given by

$$T_{f(x)}g \circ T_xf = (dg)_x \circ T_xf.$$

The verifications are all easy, and even trivial, except possibly for **Property 2.**. We shall give the proof of **Property 2** in the finite dimensional case and leave the general case to the reader.

For a form of degree 1, say

$$\omega(y) = g(y)\, dy_1,$$

with $y_1 = f_1(x)$, we find

$$(f^* \, d\omega)(x) = \big(g'(f(x)) \circ f'(x)\big) \wedge df_1(x).$$

Using the fact that $ddf_1 = 0$, together with Proposition 6 we get

$$(df^*\omega)(x) = \big(d(g \circ f)\big)(x) \wedge df_1(x),$$

which is equal to the preceding expression. Any 1-form can be expressed as a linear combination of form $g_i \, dy_i$, so that our assertion is proved for forms of degree 1.

The general formula can now be proved by induction. Using the linearity of f^*, we may assume that ω is expressed as $\omega = \psi \wedge \eta$ where ψ, η have lower degree. We apply Proposition 6 and **Property 1** to

$$f^* \, d\omega = f^*(d\psi \wedge \eta) + (-1)^r f^*(\psi \wedge d\eta)$$

and we see at once that this is equal to $df^*\omega$, because by induction, $f^* \, d\psi = df^*\psi$ and $f^* \, d\eta = df^*\eta$. This proves **Property 2**.

Example 1. Let $y_1, \ldots, y_m$ be the coordinates on V, and let μ_j be the jth coordinate function, $j = 1, \ldots, m$, so that $y_j = \mu_j(y_1, \ldots, y_m)$. Let

$$f: U \to V$$

be the map with coordinate functions

$$y_j = f_j(x) = \mu_j \circ f(x).$$

If

$$\omega(y) = g(y)\, dy_{j_1} \wedge \cdots \wedge dy_{j_s}$$

is a differential form on V, then

$$\boxed{f^*\omega = (g \circ f)\, df_{j_1} \wedge \cdots \wedge df_{j_s}.}$$

Indeed, we have for $x \in U$:

$$(f^*\omega)(x) = g(f(x))(\mu_{j_1} \circ f'(x)) \wedge \cdots \wedge (\mu_{j_s} \circ f'(x))$$

and

$$f_j'(x) = (\mu_j \circ f)'(x) = \mu_j \circ f'(x) = df_j(x).$$

Example 2. Let $f: [a, b] \to \mathbf{R}^2$ be a map from an interval into the plane, and let x, y be the coordinates of the plane. Let t be the coordinate in $[a, b]$. A differential form in the plane can be written in the form

$$\omega(x, y) = g(x, y) \, dx + h(x, y) \, dy$$

where g, h are functions. Then by definition,

$$f^*\omega(t) = g(x(t), y(t)) \frac{dx}{dt} \, dt + h(x(t), y(t)) \frac{dy}{dt} \, dt,$$

if we write $f(t) = (x(t), y(t))$. Let $G = (g, h)$ be the vector field whose components are g and h. Then we can write

$$f^*\omega(t) = G(f(t)) \cdot f'(t) \, dt,$$

which is essentially the expression which is integrated when defining the integral of a vector field along a curve.

Example 3. Let U, V be both open sets in n-space, and let $f: U \to V$ be a C^p map. If

$$\omega(y) = g(y) \, dy_1 \wedge \cdots \wedge dy_n,$$

where $y_j = f_j(x)$ is the jth coordinate of y, then

$$dy_j = D_1 f_j(x) \, dx_1 + \cdots + D_n f_j(x) \, dx_n$$

$$= \frac{\partial y_j}{\partial x_1} \, dx_1 + \cdots + \frac{\partial y_j}{\partial x_n} \, dx_n,$$

and consequently, expanding out the alternating product according to the usual multilinear and alternating rules, we find that

$$f^*\omega(x) = g(f(x)) \Delta_f(x) \, dx_1 \wedge \cdots \wedge dx_n,$$

where Δ_f is the determinant of the Jacobian matrix of f.

§4. The canonical 2-form

Consider the functor $\mathbf{E} \mapsto L(\mathbf{E})$ (continuous linear forms). If $E \to X$ is a vector bundle, then $L(E)$ will be called the **dual bundle**, and will be denoted by $\mathbf{E}^*$. For each $x \in X$, the fiber of the dual bundle is simply $L(E_x)$.

If $E = T(X)$ is the tangent bundle, then its dual is denoted by $T^*(X)$ and is called the **cotangent bundle**. Its elements are called **cotangent vectors**. The fiber of $T^*(X)$ over a point x of X is denoted by $T_x^*(X)$. For each $x \in X$ we have a pairing

$$T_x^* \times T_x \to \mathbf{R}$$

given by

$$\langle v, u \rangle$$

for $v \in T_x^*$ and $u \in T_x$ (it is the value of the linear form v at u).

We shall now describe how to construct a canonical 1-form on the cotangent bundle $T^*(X)$. For each $v \in T^*(X)$ we must define a 1-form on $T_v\big(T^*(X)\big)$.

Let $\pi \colon T^*(X) \to X$ be the canonical projection. Then the induced tangent map

$$T\pi = \pi_* \colon T\big(T^*(X)\big) \to T(X)$$

can be applied to an element z of $T_v\big(T^*(X)\big)$ and one sees at once that $\pi_* z$ lies in $T_x(X)$ if v lies in $T_x^*(X)$. Thus we can take the pairing

$$\langle v, \pi_* z \rangle = \omega_v(z)$$

to define a map (which is obviously continuous linear):

$$\omega_v \colon T_v\big(T^*(X)\big) \to \mathbf{R}.$$

Proposition 7. *This map defines a 1-form on $T^*(X)$. Let $X = U$ be open in $\mathbf{E}$ and*

$$T^*(U) = U \times \mathbf{E}^*, \qquad T\big(T^*(U)\big) = (U \times \mathbf{E}^*) \times (\mathbf{E} \times \mathbf{E}^*).$$

If $(x, v) \in U \times \mathbf{E}^*$ and $(y, w) \in \mathbf{E} \times \mathbf{E}^*$, then the local representation $\omega_{(x,v)}$ is given by

$$\langle \omega_{(x,v)}, (y, w) \rangle = \langle v, y \rangle.$$

Proof. We observe that the projection $\pi \colon U \times \mathbf{E}^* \to U$ is linear, and hence that its derivative at each point is constant, equal to the projection on the first factor. Our formula is then an immediate consequence of the definition. The local formula shows that ω is in fact a 1-form locally, and therefore globally since it has an invariant description.

Our 1-form is called the **canonical 1-form on the cotangent bundle.** Its exterior derivative $d\omega = \Omega$ will be called the **canonical 2-form.** The next proposition describes it locally.

Proposition 8. *Let U be open in $\mathbf{E}$. Let Ω be the local representation of the canonical 2-form on $T(T^*(U))$. Let $(x, v) \in U \times \mathbf{E}^*$. Let (y_1, w_1) and (y_2, w_2) be elements of $\mathbf{E} \times \mathbf{E}^*$. Then*

$$\langle \Omega_{(x,v)}, (y_1, w_1) \times (y_2, w_2) \rangle = \langle w_1, y_2 \rangle - \langle w_2, y_1 \rangle.$$

Proof. We observe that ω is linear, and thus that ω' is constant. We then apply the local formula for the exterior derivative, given in Proposition 5 of the preceding section. Our assertion becomes obvious.

§5. *The Poincaré lemma*

If ω is a differential form on a manifold and is such that $d\omega = 0$, then it is customary to say that it is **closed.** If there exists a form ψ such that $\omega = d\psi$, then one says that ω is **exact.** We shall now prove that locally, every closed form is exact.

Poincaré lemma. *Let U be an open ball in $\mathbf{E}$ and let ω be a differential form of degree ≥ 1 on U such that $d\omega = 0$. Then there exists a differential form ψ on U such that $d\psi = \omega$.*

Proof. We shall construct a linear map k from the r-forms to the $(r - 1)$-forms $(r \geq 1)$ such that

$$dk + kd = id.$$

From this relation, it will follow that whenever $d\omega = 0$, then

$$dk\omega = \omega,$$

thereby proving our proposition. We may assume that the center of the ball is the origin. If ω is an r-form, then we define $k\omega$ by the formula

$$\langle (k\omega)_x, v_1 \times \cdots \times v_{r-1} \rangle = \int_0^1 t^{r-1} \langle \omega(tx), x \times v_1 \times \cdots \times v_{r-1} \rangle \, dt.$$

We can assume that we deal with local representations and that $v_i \in \mathbf{E}$. We have

$$\langle (dk\omega)_x, v_1 \times \cdots \times v_r \rangle$$

$$= \sum_{i=1}^{r} (-1)^{i+1} \langle (k\omega)'(x)v_i, v_1 \times \cdots \times \hat{v}_i \times \cdots \times v_r \rangle$$

$$= \sum (-1)^{i+1} \int_0^1 t \langle {}^{r-1}\omega(tx), v_i \times v_1 \times \cdots \times \hat{v}_i \times \cdots \times v_r \rangle \, dt$$

$$+ \sum (-1)^{i+1} \int_0^1 t^r \langle \omega'(tx)v_i, x \times v_1 \times \cdots \times \hat{v}_i \times \cdots \times v_r \rangle \, dt.$$

On the other hand, we also have

$$\langle (kd\omega)(x), v_1 \times \cdots \times v_r \rangle = \int_0^1 t^r \langle d\omega(x), x \times v_1 \times \cdots \times v_r \rangle \, dt$$

$$= \int_0^1 t^r \langle \omega'(tx)x, v_1 \times \cdots \times v_r \rangle \, dt$$

$$+ \sum (-1)^i \int_0^1 t^r \langle \omega'(tx)v_i, x \times v_1 \times \cdots \times \hat{v}_i \times \cdots \times v_r \rangle \, dt.$$

We observe that the second terms in the expressions for $kd\omega$ and $dk\omega$ occur with opposite signs and cancel when we take the sum. As to the first terms, if we shift v_i to the ith place in the expression for $dk\omega$, then we get an extra coefficient of $(-1)^{i+1}$. Thus

$$dk\omega + kd\omega = \int_0^1 rt^{r-1} \langle \omega(tx), v_1 \times \cdots \times v_r \rangle \, dt$$

$$+ \int_0^1 t^r \langle \omega'(tx)x, v_1 \times \cdots \times v_r \rangle \, dt.$$

This last integral is simply the integral of the derivative with respect to t of

$$\langle t^r \omega(tx), v_1 \times \cdots \times v_r \rangle.$$

Evaluating this expression between $t = 0$ and $t = 1$ yields

$$\langle \omega(x), v_1 \times \cdots \times v_r \rangle$$

which proves our proposition.

We observe that we could have taken our open set U to be star-shaped instead of an open ball.

§6. Contractions and Lie derivative

Let ξ be a vector field and let ω be an r-form on a manifold X, $r \geqq 1$. Then we can define an $(r - 1)$-form $C_\xi\omega$ by the formula

$$(C_\xi\omega)(x)(v_2, \ldots, v_r) = \omega\big(\xi(x), v_2, \ldots, v_r\big),$$

for $v_2, \ldots, v_r \in T_x$. Using local representations shows at once that $C_\xi\omega$ has the appropriate order of differentiability (the minimum of ω and ξ). We call $C_\xi\omega$ the **contraction** of ω by ξ, and also denote $C_\xi\omega$ by

$$\omega \lrcorner \xi \qquad \text{or} \qquad \omega \circ \xi.$$

If f is a function, we define $C_\xi f = 0$. This operation of contraction satisfies the following properties.

CON 1. $C_\xi \circ C_\xi = 0.$

CON 2. *The association $(\xi, \omega) \mapsto C_\xi\omega = \omega \lrcorner \xi$ is bilinear. It is in fact bilinear with respect to functions, that is if φ is a function, then*

$$C_{\varphi\xi} = \varphi C_\xi \qquad \text{and} \qquad C_\xi(\varphi\omega) = \varphi C_\xi\omega.$$

CON 3. *If ω, ψ are differential forms and $r = \deg \omega$, then*

$$C_\xi(\omega \wedge \psi) = (C_\xi\omega) \wedge \psi + (-1)^r \omega \wedge C_\xi\psi.$$

These three properties follow at once from the definitions.

Example. Let $X = \mathbf{R}^n$, and let

$$\omega(x) = dx_1 \wedge \cdots \wedge dx_n.$$

If ξ is a vector field on $\mathbf{R}^n$, then we have the local representation

$$(\omega \lrcorner \xi)(x) = \sum_{i=1}^{n} (-1)^{i+1}\xi_i(x)\, dx_1 \wedge \cdots \wedge \widehat{dx_i} \wedge \cdots \wedge dx_n.$$

We also have immediately from the definition of the exterior derivative,

$$d(\omega \lrcorner \xi) = \sum_{i=1}^{n} \frac{\partial \xi_i(x)}{\partial x_i}\, dx_1 \wedge \cdots \wedge dx_n,$$

letting $\xi = (\xi_1, \ldots, \xi_n)$ in terms of its components ξ_i.

We can define the **Lie derivative** of an r-form as we did before for vector fields. Namely, we shall evaluate the following limit:

$$(\mathscr{L}_\xi \omega)(x) = \lim_{t \to 0} \frac{1}{t} [(\alpha_t^* \omega)(x) - \omega(x)],$$

or in other words,

$$\mathscr{L}_\xi \omega = \frac{d}{dt} (\alpha_t^* \omega) \Big|_{t=0}$$

where α is the flow of the vector field ξ, and we call $\mathscr{L}_\xi$ the **Lie derivative** again, applied to the differential form ω.

Proposition 9. *Let ξ be a vector field and ω a differential form of degree $r \geq 1$. If $\xi_1, \ldots, \xi_r$ are vector fields, then the Lie derivative $\mathscr{L}_\xi \omega$ is given by*

$$\mathscr{L}_\xi \omega = \frac{d}{dt} \langle \omega \circ \alpha_t, \; \alpha_{t*}\xi_1 \times \cdots \times \alpha_{t*}\xi_r \rangle \Big|_{t=0}.$$

If ξ, ξ_i, ω denote the local representations of the vector fields and the form respectively, then the Lie derivative $\mathscr{L}_\xi \omega$ has the local representation

$$\langle \omega'(x)\xi(x), \; \xi_1(x) \times \cdots \times \xi_r(x) \rangle = \langle (\mathscr{L}_\xi \omega)(x), \; \xi_1(x) \times \cdots \times \xi_r(x) \rangle$$

$$+ \sum_{i=1}^{r} \langle \omega(x), \; \xi_1(x) \times \cdots \times \xi'(x)\xi_i(x) \times \cdots \times \xi_r(x) \rangle.$$

Proof. The proof is routine using the definitions. The first assertion is obvious by the definition of the pull-back of a form. For the local representation, let

$$F(t) = \langle (\alpha_t^* \omega)(x), \; \xi_1(x) \times \cdots \times \xi_r(x) \rangle$$

$$= \langle \omega(\alpha(t, x)), \; D_2\alpha(t, x)\xi_1(x) \times \cdots \times D_2\alpha(t, x)\xi_r(x) \rangle.$$

Then the Lie derivative $(\mathscr{L}_\xi \omega)(x)$ is precisely $F'(0)$, and we have

$$F'(t) = \left\langle \frac{d}{dt} \alpha_t^* \omega(x), \; \xi_1(x) \times \cdots \times \xi_r(x) \right\rangle.$$

To compute $F'(t)$ we use the rule for the derivative of a product, and get

$$F'(t) =$$

$$\langle \omega'(\alpha(t, x))D_1\alpha(t, x), \; D_2\alpha(t, x)\xi_1(x) \times \cdots \times D_2\alpha(t, x)\xi_r(x) \rangle +$$

$$\sum_{i=1}^{r} \langle \omega(\alpha(t, x)), \; D_2\alpha(t, x)\xi_1(x) \times \cdots \times D_1D_2\alpha(t, x)\xi_i(x) \times \cdots \times D_2\alpha(t, x)\xi_r(x) \rangle.$$

Putting $t = 0$ and using the differential equation satisfied by $D_2\alpha(t, x)$, we get precisely the local expression as stated in the proposition. Remember the initial condition $D_2\alpha(0, x) = id$.

Proposition 10. *As a map on differential forms, the Lie derivative satisfies the following properties.*

LIE 1. $\mathscr{L}_\xi = d \circ C_\xi + C_\xi \circ d.$

LIE 2. $(\mathscr{L}_\xi \omega \wedge \psi) = \mathscr{L}_\xi \omega \wedge \psi + \omega \wedge \mathscr{L}_\xi \psi.$

Proof. Let $\xi_1, \ldots, \xi_r$ be vector fields, and ω an r-form. Using the definition of the contraction and the local formula of Proposition 5, we find that $C_\xi \, d\omega$ is given locally by

$$\langle C_\xi \, d\omega(x), \xi_1(x) \times \cdots \times \xi_r(x)\rangle = \langle \omega'(x)\xi(x), \xi_1(x) \times \cdots \times \xi_r(x)\rangle$$

$$+ \sum_{i=1}^{r} (-1)^i \langle \omega'(x)\xi_i(x), \xi(x) \times \xi_1(x) \times \cdots \times \widehat{\xi_i(x)} \times \cdots \times \xi_r(x)\rangle.$$

On the other hand, $dC_\xi\omega$ is given by

$$\langle dC_\xi\omega(x), \xi_1(x) \times \cdots \times \xi_r(x)\rangle$$

$$= \sum_{i=1}^{r} (-1)^{i+1}\langle (C_\xi\omega)'(x)\xi_i(x), \xi_1(x) \times \cdots \times \widehat{\xi_i(x)} \times \cdots \times \xi_r(x)\rangle.$$

To compute $(C_\xi\omega)'(x)$ is easy, going back to the definition of the derivative. At vectors $v_1, \ldots, v_{r-1}$, the form $C_\xi\omega(x)$ has the value

$$\langle \omega(x), \xi(x) \times v_1 \times \cdots \times v_{r-1}\rangle.$$

Differentiating this last expression with respect to x and evaluating at a vector h we get

$$\langle \omega'(x)h, \xi(x) \times v_1 \times \cdots \times v_{r-1}\rangle + \langle \omega(x), \xi'(x)h \times v_1 \times \cdots \times v_{r-1}\rangle.$$

Hence $\langle dC_\xi\omega(x), \xi_1(x) \times \cdots \times \xi_r(x)\rangle$ is equal to

$$\sum_{i=1}^{r} (-1)^{i+1}\langle \omega'(x)\xi_i(x), \xi(x) \times \xi_1(x) \times \cdots \times \widehat{\xi_i(x)} \times \cdots \times \xi_r(x)\rangle$$

$$+ \sum_{i=1}^{r} (-1)^{i+1}\langle \omega(x), \xi'(x)\xi_i(x) \times \xi_1(x) \times \cdots \times \widehat{\xi_i(x)} \times \cdots \times \xi_r(x)\rangle.$$

Shifting $\xi'(x)\xi_i(x)$ to the ith place in the second sum contributes a sign of $(-1)^{i-1}$ which gives 1 when multiplied by $(-1)^{i+1}$. Adding the two local representations for $dC_\xi \omega$ and $C_\xi \, d\omega$, we find precisely the expression of Proposition 9, thus proving **LIE 1**.

As for **LIE 2**, it consists in using the multiplication rule for d and C_ξ in Proposition 6, **EXD 1**, and **CON 3**. The corresponding rule for $\mathscr{L}_\xi$ follows at once. (Terms will cancel just the right way.)

In the two preceding propositions, we deal with a time-independent vector field. It is occasionally useful to deal with a time-dependent vector field. Furthermore, the expression for the derivative $F'(t)$ in Proposition 9, together with property **LIE 1**, are actually special cases of a more general formula as follows.

Proposition 11. *Let ξ_t be a time-dependent vector field, α its flow, and let ω be a differential form. Then*

$$\frac{d}{dt}(\alpha_t^* \omega) = \alpha_t^*(d \circ C_{\xi_t}\omega + C_{\xi_t} \circ d\omega) = \alpha_t^*(\mathscr{L}_{\xi_t}\omega).$$

Proof. The proof follows the same lines as before.

We note that Proposition 10 is actually a corollary of Proposition 11, by putting $t = 0$ and observing that $\alpha_0^* = id$.

§7. Darboux theorem

Let **E** be a Banach space and let

$$\omega : \mathbf{E} \times \mathbf{E} \to \mathbf{R}$$

be a continuous bilinear map. Then ω induces a continuous linear map

$$\lambda_\omega = \lambda : \mathbf{E} \to \mathbf{E}^*$$

which to each $v \in \mathbf{E}$ associates the functional $\lambda(v)$ such that

$$\lambda(v)(w) = \omega(v, w).$$

We have a similar map on the other side. If both these mappings are top-linear isomorphisms of **E** and **E*** then we say that ω is **non-singular**. If such a non-singular form exists, then we say that **E** is **self-dual**. For instance, a Hilbert space is self-dual.

If **E** is finite dimensional, it suffices for a bilinear form to be non-singular that its kernels on the right and on the left be 0. (The kernels are the kernels of the associated maps λ as above.) However, in the infinite dimensional case, this condition on the kernels is not sufficient any more.

If $\mathbf{E} = \mathbf{R}^n$ then the usual scalar product establishes the self-duality of $\mathbf{R}^n$. This self-duality arises from other forms, and in this section we are especially interested in the self-duality arising from alternating forms. If $\mathbf{E}$ is finite dimensional and ω is an element of $L_a^2(\mathbf{E})$, that is an alternating 2-form, which is non-singular, then one sees easily that the dimension of $\mathbf{E}$ is even.

Example. An example of such a form on $\mathbf{R}^{2n}$ is the following. Let

$$v = (v_1, \ldots, v_n, v_1', \ldots, v_n')$$

$$w = (w_1, \ldots, w_n, w_1', \ldots, w_n')$$

be elements of $\mathbf{R}^{2n}$, with components v_i, v_i', w_i, w_i'. Letting

$$\omega(v, w) = \sum_{i=1}^n (v_i w_i' - v_i' w_i)$$

defines a non-singular 2-form ω on $\mathbf{R}^{2n}$. It is an exercise of linear algebra to prove that any non-singular 2-form on $\mathbf{R}^{2n}$ is linearly isomorphic to this particular one in the following sense. If

$$f \colon E \to F$$

is a linear isomorphism between two finite dimensional spaces, then it induces an isomorphism

$$f^* \colon L_a^2(F) \to L_a^2(E).$$

We call forms ω on E and ψ on F **linearly isomorphic** if there exists a linear isomorphism f such that $f^*\psi = \omega$. Thus up to a linear isomorphism, there is only one non-singular 2-form on $\mathbf{R}^{2n}$. (For a proof, cf. for instance my book *Algebra*.)

We are interested in the same question on a manifold locally. Let U be open in the Banach space $\mathbf{E}$ and let $x_0 \in U$. A 2-form

$$\omega \colon U \to L_a^2(\mathbf{E})$$

is said to be **non-singular** if each form $\omega(x)$ is non-singular. If ξ is a vector field on U, then $\xi \,\lrcorner\, \omega = \omega \circ \xi$ is a 1-form, whose value at (x, w) is given by

$$(\omega \circ \xi)(x)(w) = \omega(x)(\xi(x), w).$$

Proposition 12. *Let ω be a non-singular 2-form on an open set U in $\mathbf{E}$. The association*

$$\xi \mapsto \omega \circ \xi$$

is a linear isomorphism between the space of vector fields on U and the space of 1-forms on U.

Proof. Let $(v, w) \mapsto \langle v, w \rangle$ be a fixed non-singular continuous bilinear form on $\mathbf{E}$. For each $x \in U$ there exists a unique linear map

$$A_x : \mathbf{E} \to \mathbf{E}$$

such that

$$\omega(x)(v, w) = \langle A_x v, w \rangle,$$

and it is immediately verified that A_x is in fact continuous. Since $\omega(x)$ is assumed non-singular, it follows that A_x is a toplinear isomorphism. The map

$$x \mapsto A_x$$

is therefore seen to be a morphism of U into Laut $(\mathbf{E})$.

If θ is a 1-form on U, then for each x there exists an element $\eta(x) \in \mathbf{E}$ such that

$$\theta(x)(w) = \langle \eta(x), w \rangle, \qquad \text{all } w \in \mathbf{E},$$

and it is immediately seen that $\eta : U \to \mathbf{E}$ is a morphism (composition of the map

$$\theta : U \to \mathbf{E}^*$$

and the toplinear isomorphism between $\mathbf{E}^*$ and $\mathbf{E}$ given by the non-singular form $\langle \, , \, \rangle$.) By the non-singularity of $\omega(x)$, there exists an element $\xi(x) \in \mathbf{E}$ such that

$$\omega(x)(\xi(x), w) = \theta(x)(w) = \langle \eta(x), w \rangle, \qquad \text{all } w \in \mathbf{E}.$$

We see that

$$\xi(x) = A_x^{-1} \eta(x),$$

whence ξ is also a morphism. Thus the map

$$\xi \mapsto \omega \circ \xi$$

is a surjective linear map from the space of vector fields to the space of 1-forms, and it is obviously injective, thus proving our proposition.

Let

$$\omega : U \to L_a^2(U)$$

be a 2-form on an open set U in $\mathbf{E}$. If there exists a local isomorphism f at a point $x_0 \in U$, say

$$f: U_1 \to V_1,$$

and a 2-form ψ on V_1 such that $f^*\psi = \omega$ (or more accurately, ω restricted to U_1), then we say that ω is **locally isomorphic** to ψ at x_0. Observe that in the case of an isomorphism we can take a direct image of forms, and we shall also write

$$f_*\omega = \psi$$

instead of $\omega = f^*\psi$. In other words, $f_* = (f^{-1})^*$.

Example. On $\mathbf{R}^{2n}$ we have the constant form of the previous example. In terms of local coordinates $(x_1, \ldots, x_n, y_1, \ldots, y_n)$, this form has the local expression

$$\omega(x, y) = \sum_{i=1}^{n} dx_i \wedge dy_i.$$

This 2-form will be called the **standard 2-form on $\mathbf{R}^{2n}$**.

The Darboux theorem states that any non-singular closed 2-form in $\mathbf{R}^{2n}$ is locally isomorphic to the standard form, that is that in a suitable chart at a point, it has the standard expression of the above example. A technique to show that certain forms are isomorphic was used by Moser [20], who pointed out that his arguments also prove the classical Darboux theorem. Alan Weinstein observed that the proof applies to the infinite dimensional case, whose statement is as follows.

Darboux theorem. *Let $\mathbf{E}$ be a self-dual Banach space. Let*

$$\omega: U \to L_a^2(\mathbf{E})$$

be a non-singular closed 2-form on an open set of $\mathbf{E}$, and let $x_0 \in U$. Then ω is locally isomorphic at x_0 to the constant form $\omega(x_0)$.

Proof. Let $\omega_0 = \omega(x_0)$, and let

$$\omega_t = \omega_0 + t(\omega - \omega_0), \qquad\qquad 0 \leq t \leq 1.$$

We wish to find a time-dependent vector field ξ_t locally at 0 such that if α denotes its flow, then

$$\alpha_t^* \omega_t = \omega_0.$$

Then the local isomorphism α_1 satisfies the requirements of the theorem. By the Poincaré lemma, there exists a 1-form θ locally at 0 such that

$$\omega - \omega_0 = d\theta,$$

and without loss of generality, we may assume that $\theta(x_0) = 0$. We contend that the time-dependent vector field ξ_t, such that

$$\omega_t \circ \xi_t = -\theta,$$

has the desired property. Let α be its flow. If we shrink the domain of the vector field near x_0 sufficiently, and use the fact that $\theta(x_0) = 0$, then we can use the local existence theorem (Proposition 1 of Chapter IV, §1) to see that the flow can be integrated at least to $t = 1$ for all points x in this small domain. We shall now verify that

$$\frac{d}{dt}(\alpha_t^* \omega_t) = 0.$$

This will prove that $\alpha_t^* \omega_t$ is constant. Since we have $\alpha_0^* \omega_0 = \omega_0$ because

$$\alpha(0, x) = x \qquad \text{and} \qquad D_2\alpha(0, x) = id,$$

it will conclude the proof of the theorem.

We compute locally. We use the local formula of Proposition 11, §6, and formula **LIE 1.** which reduces to

$$\mathscr{L}_{\xi_t}\omega_t = d(\omega_t \circ \xi_t),$$

because $d\omega_t = 0$. We find

$$\frac{d}{dt}(\alpha_t^* \omega_t) = \alpha_t^* \left(\frac{d}{dt}\omega_t\right) + \alpha_t^*(\mathscr{L}_{\xi_t}\omega_t)$$

$$= \alpha_t^* \left(\frac{d}{dt}\omega_t + d(\omega_t \circ \xi_t)\right)$$

$$= \alpha_t^*(\omega - \omega_0 - d\theta)$$

$$= 0.$$

This proves Darboux's theorem.

Remark 1. The proof of the Poincaré lemma can also be cast in the above style. For instance, let $\phi_t(x) = tx$ be a retraction of an open set around 0.

Let ξ_t be the vector field whose flow is ϕ_t, and let ω be a closed form. Then

$$\frac{d}{dt}\phi_t^*\omega = \phi_t^*\mathcal{L}_{\xi_t}\omega = \phi_t^*\, dC_{\xi_t}\omega = d\phi_t^* C_{\xi_t}\omega.$$

Since $\phi_0^*\omega = 0$ and ϕ_1 is the identity, we see that

$$\omega = \phi_1^*\omega - \phi_0^*\omega = \int_0^1 \frac{d}{dt}\phi_t^*\omega\, dt = d\int_0^1 \phi_t^* C_{\xi_t}\omega\, dt$$

is exact.

Remark 2. A manifold with a nonsingular closed 2-form is sometimes called a **symplectic** manifold. Such manifolds occur a lot in mechanics.

CHAPTER VI

The Theorem of Frobenius

Having acquired the language of vector fields, we return to differential equations and give a generalization of the local existence theorem known as the Frobenius theorem, whose proof will be reduced to the standard case discussed in Chapter IV. We state the theorem in §1. The reader should note that he needs only to know the definition of the bracket of two vector fields in order to understand the proof. It is convenient to insert also a formulation in terms of differential forms, for which the reader needs to know the local definition of the exterior derivative. However, the condition involving differential forms is proved to be equivalent to the vector field condition at the very beginning, and does not reappear explicitly afterwards.

We shall follow essentially the proof given by Dieudonné in his *Foundations of Modern Analysis*, allowing for the fact that we use freely the geometric language of vector bundles, which is easier to grasp.

It is convenient to recall in §2 the statements concerning the existence theorems for differential equations depending on parameters. The proof of the Frobenius theorem proper is given in §3. An important application to Lie groups is given in §5, after formulating the theorem globally.

The present chapter is logically independent of the next one on Riemannian metrics.

§1. Statement of the theorem

Let X be a manifold of class C^p ($p \geqq 2$). A subbundle E of its tangent bundle will also be called a **tangent subbundle** over X. We contend that the following two conditions concerning such a subbundle are equivalent.

FR 1. *For each point $z \in X$ and vector fields ξ, η at z (i.e. defined on an open neighborhood of z) which lie in E (i.e. such that the image of each point of X under ξ, η lies in E), the bracket $[\xi, \eta]$ also lies in E.*

FR 2. *For each point $z \in X$ and differential form ω of degree 1 at z which vanishes on E, the form $d\omega$ vanishes on $\xi \times \eta$ whenever ξ, η are two vector fields at z which lie in E.*

135

The equivalence is essentially a triviality. Indeed, assume **FR 1**. Let ω vanish on E. Then

$$\langle d\omega, \xi \times \eta \rangle = -\langle \omega, [\xi, \eta] \rangle - \eta \langle \omega, \xi \rangle + \xi \langle \omega, \eta \rangle.$$

By assumption the right-hand side is 0 when evaluated at z. Conversely, assume **FR 2**. Let ξ, η be two vector fields at z lying in E. If $[\xi, \eta](z)$ is not in E, then we see immediately from a local product representation and the Hahn-Banach theorem that there exists a differential form ω of degree 1 defined on a neighborhood of z which is 0 on E_z and non-zero on $[\xi, \eta](z)$, thereby contradicting the above formula.

We shall now give a third condition equivalent to the above two, and actually, we shall not refer to **FR 2** any more. We remark merely that in the finite dimensional case, it is easy to prove that when a differential form ω satisfies condition **FR 2**, then $d\omega$ can be expressed locally in a neighborhood of each point as a finite sum

$$d\omega = \sum \gamma_i \wedge \omega_i$$

where γ_i and ω_i are of degree 1 and each ω_i vanishes on E. We leave this as an exercise to the reader.

Let E be a tangent subbundle over X. We shall say that E is **integrable** at a point x_0 if there exists a submanifold Y of X containing x_0 such that the tangent map of the inclusion

$$j: Y \to X$$

induces a VB-isomorphism of TY with the subbundle E. Equivalently, we could say that for each point $y \in Y$, the tangent map

$$T_y j: T_y Y \to T_y X$$

induces a toplinear isomorphism of $T_y Y$ on E_y. Note that our condition defining integrability is local at x_0. We say that E is **integrable** if it is integrable at every point.

Using the functoriality of vector fields, and their relations under tangent maps and the bracket product, we see at once that if E is integrable, then it satisfies **FR 1**. Indeed, locally, vector fields having their values in E are related to vector fields over Y under the inclusion mapping.

Frobenius' theorem asserts the converse.

Theorem 1. *Let X be a manifold of class C^p ($p \geq 2$) and let E be a tangent subbundle over X. Then E is integrable if and only if E satisfies condition* **FR 1**.

The proof of Frobenius' theorem will be carried out by analyzing the situation locally and reducing it to the standard theorem for ordinary differential equations. Thus we now analyze the condition **FR 1** in terms of its local representation.

Suppose that we work locally, over a product $U \times V$ of open subsets of Banach spaces **E** and **F**. Then the tangent bundle $T(U \times V)$ can be written in a natural way as a direct sum. Indeed, for each point (x, y) in $U \times V$ we have

$$T_{(x,y)}(U \times V) = T_x(U) \times T_y(V).$$

One sees at once that the collection of fibers $T_x(U) \times 0$ (contained in $T_x(U) \times T_y(V)$) forms a subbundle which will be denoted by $T_1(U \times V)$ and will be called the **first factor** of the tangent bundle. One could define $T_2(U \times V)$ similarly, and

$$T(U \times V) = T_1(U \times V) \oplus T_2(U \times V).$$

A subbundle E of $T(X)$ is integrable at a point $z \in X$ if and only if there exists an open neighborhood W of z and an isomorphism

$$\varphi: U \times V \to W$$

of a product onto W such that the composition of maps

$$T_1(U \times V) \xrightarrow{\text{inc.}} T(U \times V) \xrightarrow{T\varphi} T(W)$$

induces a VB-isomorphism of $T_1(U \times V)$ onto $E \mid W$ (over φ). Denoting by φ_y the map of U into W given by $\varphi_y(x) = \varphi(x, y)$, we can also express the integrability condition by saying that $T_x\varphi_y$ should induce a toplinear isomorphism of **E** onto $E_{\varphi(x,y)}$ for all (x, y) in $U \times V$. We note that in terms of our local product structure, $T_x\varphi_y$ is nothing but the partial derivative $D_1\varphi(x, y)$.

Given a subbundle of $T(X)$, and a point in the base space X, we know from the definition of a subbundle in terms of a local product decomposition that we can find a product decomposition of an open neighborhood of this point, say $U \times V$, such that the point has coordinates (x_0, y_0) and such that the subbundle can be written in the form of an exact sequence

$$0 \to U \times V \times \mathbf{E} \xrightarrow{f} U \times V \times \mathbf{E} \times \mathbf{F}$$

with the map

$$f(x_0, y_0): \mathbf{E} \to \mathbf{E} \times \mathbf{F}$$

equal to the canonical embedding of $\mathbf{E}$ on $\mathbf{E} \times 0$. For a point (x, y) in $U \times V$ the map $f(x, y)$ has two components $f_1(x, y)$ and $f_2(x, y)$ into $\mathbf{E}$ and $\mathbf{F}$ respectively. Taking a suitable VB-automorphism of $U \times V \times \mathbf{E}$ if necessary, we may assume without loss of generality that $f_1(x, y)$ is the identity. We now write $f(x, y) = f_2(x, y)$. Then

$$f: U \times V \to L(\mathbf{E}, \mathbf{F})$$

is a morphism (of class C^{p-1}) which describes our subbundle completely.

We shall interpret condition **FR 1** in terms of the present situation. If

$$\xi: U \times V \to \mathbf{E} \times \mathbf{F}$$

is the local representation of a vector field over $U \times V$, we let ξ_1 and ξ_2 be its projections on $\mathbf{E}$ and $\mathbf{F}$ respectively. Then ξ lies in the image of $\tilde{f}$ if and only if

$$\xi_2(x, y) = f(x, y)\xi_1(x, y)$$

for all (x, y) in $U \times V$, or in other words, if and only if ξ is of the form

$$\xi(x, y) = \big(\xi_1(x, y), f(x, y)\xi_1(x, y)\big)$$

for some morphism (of class C^{p-1})

$$\xi_1: U \times V \to \mathbf{E}.$$

We shall also write the above condition symbolically, namely

$$\xi = (\xi_1, f \cdot \xi_1). \tag{1}$$

If ξ, η are the local representations of vector fields over $U \times V$, then the reader will verify at once from the local definition of the bracket (Proposition 3 of Chapter V, §1) that $[\xi, \eta]$ lies in the image of $\tilde{f}$ if and only if

$$Df(x, y) \cdot \xi(x, y) \cdot \eta_1(x, y) = Df(x, y) \cdot \eta(x, y) \cdot \xi_1(x, y)$$

or symbolically,

$$Df \cdot \xi \cdot \eta_1 = Df \cdot \eta \cdot \xi_1. \tag{2}$$

We have now expressed all the hypotheses of Theorem 1 in terms of local data, and the heart of the proof will consist in proving the following result.

Theorem 2. *Let U, V be open subsets of Banach spaces $\mathbf{E}, \mathbf{F}$ respectively. Let*

$$f: U \times V \to L(\mathbf{E}, \mathbf{F})$$

be a C^r-morphism $(r \geqq 1)$. *Assume that if*

$$\xi_1, \eta_1 \colon U \times V \to \mathbf{E}$$

are two morphisms, and if we let

$$\xi = (\xi_1, f \cdot \xi_1) \qquad and \qquad \eta = (\eta_1, f \cdot \eta_1)$$

then relation (2) above is satisfied. Let (x_0, y_0) be a point of $U \times V$. Then there exist open neighborhoods U_0, V_0 of x_0, y_0 respectively, contained in U, V, and a unique morphism $\alpha \colon U_0 \times V_0 \to V$ such that

$$D_1\alpha(x, y) = f\big(x, \alpha(x, y)\big)$$

and $\alpha(x_0, y) = y$ for all (x, y) in $U_0 \times V_0$.

We shall prove Theorem 2 in §3. We now indicate how Theorem 1 follows from it. We denote by α_y the map $\alpha_y(x) = \alpha(x, y)$, viewed as a map of U_0 into V. Then our differential equation can be written

$$D\alpha_y(x) = f(x, \alpha_y(x)).$$

We let

$$\varphi \colon U_0 \times V_0 \to U \times V$$

be the map $\varphi(x, y) = \big(x, \alpha_y(x)\big)$. It is obvious that $D\varphi(x_0, y_0)$ is a toplinear isomorphism, so that φ is a local isomorphism at (x_0, y_0). Furthermore, for $(u, v) \in \mathbf{E} \times \mathbf{F}$ we have

$$D_1\varphi(x, y) \cdot (u, v) = \big(u, D\alpha_y(x) \cdot u\big) = \big(u, f(x, \alpha_y(x)) \cdot u\big)$$

which shows that our subbundle is integrable.

§2. *Differential equations depending on a parameter*

Proposition 1. *Let U, V be open sets in Banach spaces $\mathbf{E}, \mathbf{F}$ respectively. Let J be an open interval of $\mathbf{R}$ containing 0, and let*

$$g \colon J \times U \times V \to \mathbf{F}$$

be a morphism of class C^r $(r \geqq 1)$. Let (x_0, y_0) be a point in $U \times V$. Then there exist open balls J_0, U_0, V_0 centered at $0, x_0, y_0$ and contained J, U, V respectively, and a unique morphism of class C^r

$$\beta \colon J_0 \times U_0 \times V_0 \to V$$

such that $\beta(0, x, y) = y$ and

$$D_1\beta(t, x, y) = g\big(t, x, \beta(t, x, y)\big)$$

for all $(t, x, y) \in J_0 \times U_0 \times V_0$.

Proof. This follows from the existence and uniqueness of local flows, by considering the ordinary vector field on $U \times V$

$$G: J \times U \times V \to \mathbf{E} \times \mathbf{F}$$

given by $G(t, x, y) = \big(0, g(t, x, y)\big)$. If $B(t, x, y)$ is the local flow for G, then we let $\beta(t, x, y)$ be the projection on the second factor of $B(t, x, y)$. The reader will verify at once that β satisfies the desired conditions. The uniqueness is clear.

Let us keep the initial condition y fixed, and write

$$\beta(t, x) = \beta(t, x, y).$$

From Chapter IV, §1 we obtain also the differential equation satisfied by β in its second variable:

Proposition 2. *Let the notation be as in Proposition 1, and with y fixed, let $\beta(t, x) = \beta(t, x, y)$. Then $D_2\beta(t, x)$ satisfies the differential equation*

$$D_1 D_2\beta(t, x) \cdot v = D_2 g\big(t, x, \beta(t, x)\big) \cdot v + D_3 g\big(t, x, \beta(t, x)\big) \cdot D_2\beta(t, x) \cdot v,$$

for every $v \in \mathbf{E}$.

Proof. Here again, we consider the vector field as in the proof of Proposition 1, and apply the formula for the differential equation satisfied by $D_2\beta$ as in Chapter IV, §1.

§3. Proof of the theorem

In the application of Proposition 1 to the proof of Theorem 2, we take our morphism g to be

$$g(t, z, y) = f(x_0 + tz, y) \cdot z$$

with z in a small ball $\mathbf{E}_0$ around the origin in $\mathbf{E}$, and y in V. It is convenient to make a translation, and without loss of generality we can assume that $x_0 = 0$ and $y_0 = 0$. From Proposition 1 we then obtain

$$\beta: J_0 \times \mathbf{E}_0 \times V_0 \to V$$

with initial condition $\beta(0, z, y) = y$ for all $z \in \mathbf{E}_0$, satisfying the differential equation

$$D_1\beta(t, z, y) = f(tz, \beta(t, z, y)) \cdot z.$$

Making a change of variables of type $t = as$ and $z = a^{-1}x$ for a small positive number a, we see at once that we may assume that J_0 contains 1, provided we take $\mathbf{E}_0$ sufficiently small. As we shall keep y fixed from now on, we omit it from the notation, and write $\beta(t, z)$ instead of $\beta(t, z, y)$. Then our differential equation is

$$D_1\beta(t, z) = f(tz, \beta(t, z)) \cdot z. \tag{3}$$

We observe that if we knew the existence of α in the statement of Theorem 2, then letting $\beta(t, z) = \alpha(x_0 + tz)$ would yield a solution of our differential equation. Thus the uniqueness of α follows. To prove its existence, we start with β and contend that the map

$$\alpha(x) = \beta(1, x)$$

has the required properties for small $|x|$. To prove our contention it will suffice to prove that

$$D_2\beta(t, z) = tf(tz, \beta(t, z)) \tag{4}$$

because if that relation holds, then

$$D\alpha(x) = D_2\beta(1, x) = f(x, \beta(1, x)) = f(x, \alpha(x))$$

which is precisely what we want.

From Proposition 2, we obtain for any vector $v \in \mathbf{E}$,

$$D_1D_2\beta(t, z) \cdot v = tD_1f(tz, \beta(t, z)) \cdot v \cdot z$$

$$+ D_2f(tz, \beta(t, z)) \cdot D_2\beta(t, z) \cdot v \cdot z + f(tz, \beta(t, z)) \cdot v$$

We now let $k(t) = D_2\beta(t, z) \cdot v - tf(tz, \beta(t, z)) \cdot v$. Then one sees at once that $k(0) = 0$ and we contend that

$$Dk(t) = D_2f(tz, \beta(t, z)) \cdot k(t) \cdot z. \tag{5}$$

We use the main hypothesis of our theorem, namely relation (2), in which we take ξ_1 and η_1 to be the fields v and z respectively. We compute Df using the formula for the partial derivatives, and apply it to this special case. Then (5) follows immediately. It is a linear differential equation satisfied by $k(t)$, and by Corollary 2 of Proposition 2 of Chapter IV, §1 we know that the solution 0 is the unique solution. Thus $k(t) = 0$ and relation (4) is proved. The theorem also.

§4. The global formulation

Let X be a manifold. Let F be a tangent subbundle. By an **integral manifold** for F, we shall mean an injective immersion

$$f: Y \to X$$

such that at every point $y \in Y$, the tangent map

$$T_y f: T_y Y \to T_{f(y)} X$$

induces a toplinear isomorphism of $T_y Y$ on the subspace $F_{f(y)}$ of $T_{f(y)} X$. Thus Tf induces locally an isomorphism of the tangent bundle of Y with the bundle F over $f(Y)$.

Observe that the image $f(Y)$ itself may not be a submanifold of X. For instance, if F has dimension 1 (i.e. the fibers of F have dimension 1), an integral manifold for F is nothing but an integral curve from the theory of differential equations, and this curve may wind around X in such a way that its image is dense. A special case of this occurs if we consider the torus as the quotient of the plane by the subgroup generated by the two unit vectors. A straight line with irrational slope in the plane gets mapped on a dense integral curve on the torus.

If Y is a submanifold of X, then of course the inclusion $j: Y \to X$ is an injective immersion, and in this case, the condition that it be an integral manifold for F simply means that $T(Y) = F \mid Y$ (F restricted to Y).

We now have the local uniqueness of integral manifolds, corresponding to the local uniqueness of integral curves.

Theorem 3. *Let Y, Z be integral submanifolds of X for the subbundle F of TX, passing through a point x_0. Then there exists an open neighborhood U of x_0 in X, such that*

$$Y \cap U = Z \cap U.$$

Proof. Let U be an open neighborhood of x_0 in X such that we have a chart

$$U \to V \times W$$

with

$$x_0 \mapsto (y_0, w_0),$$

and Y corresponds to all points (y, w_0), $y \in V$. In other words, Y corresponds to a factor in the product in the chart. If V is open in $\mathbf{F}_1$ and W open in

$\mathbf{F}_2$, with $\mathbf{F}_1 \times \mathbf{F}_2 = \mathbf{E}$, then the subbundle $\mathbf{F}$ is represented by the projection

$$V \times W \times \mathbf{F}_1$$
$$\downarrow$$
$$V \times W$$

Shrinking Z, we may assume that $Z \subset U$. Let $h: Z \to V \times W$ be the restriction of the chart to Z, and let $h = (h_1, h_2)$ be represented by its two components. By assumption, $h'(x)$ maps $\mathbf{E}$ into $\mathbf{F}_1$ for every $x \in Z$. Hence h_2 is constant, so that $h(Z)$ is contained in the factor $V \times \{w_0\}$. It follows at once that $h(Z) = V_1 \times \{w_0\}$ for some open V_1 in V, and we can shrink U to a product $V_1 \times W_1$ (where W_1 is a small open set in W containing w_0) to conclude the proof.

We wish to get a maximal connected integral manifold for an integrable subbundle F of TX passing through a given point, just as we obtained a maximal integral curve. For this, it is just as easy to deal with the non-connected case, following Chevalley's treatment in his book on *Lie Groups*. (Note the historical curiosity that vector bundles were invented about a year after Chevalley published his book, so that the language of vector bundles, or the tangent bundle, is absent from Chevalley's presentation. In fact, Chevalley had to use a terminology which now appears terribly confusing for the notion of a tangent subbundle, and it will not be repeated here!)

We give a new manifold structure to X, depending on the integrable tangent subbundle F, and the manifold thus obtained will be denoted by X_F. This manifold has the same set of points as X. Let $x \in X$. We know from the local uniqueness theorem that a submanifold Y of X which is at the same time an integral manifold for F is locally uniquely determined. A chart for this submanifold locally at x is taken to be a chart for X_F. It is immediately verified that the collection of such charts is an atlas, which defines our manifold X_F. (We lose one order of differentiability.) The identity mapping

$$j: X_F \to X$$

is then obviously an injective immersion, satisfying the following universal properties.

Theorem 4. *Let F be an integrable tangent subbundle over X. If*

$$f: Y \to X$$

is a morphism such that $Tf\colon TY \to TX$ *maps* TY *into* F, *then the induced map*

$$f_F\colon Y \to X_F$$

(same values as f *but viewed as a map into the new manifold* X_F*) is also a morphism. Furthermore, if* f *is an injective immersion, then* f_F *induces an isomorphism of* Y *onto an open subset of* X_F.

Proof. Using the local product structure as in the proof of the local uniqueness Theorem 3, we see at once that f_F is a morphism. In other words, locally, f maps a neighborhood of each point of Y into a submanifold of X which is tangent to F. If in addition f is an injective immersion, then from the definition of the charts on X_F, we see that f_F maps Y bijectively onto an open subset of X_F, and is a local isomorphism at each point. Hence f_F induces an isomorphism of Y with an open subset of X_F, as was to be shown.

Corollary. *Let* $X_F(x_0)$ *be the connected component of* X_F *containing a point* x_0. *If* $f\colon Y \to X$ *is an integral manifold for* F *passing through* x_0, *and* Y *is connected, then there exists a unique morphism* $h\colon Y \to X_F(x_0)$ *making the following diagram commutative,*

$$
\begin{array}{ccc}
Y & \xrightarrow{\ h\ } & X_F(x_0) \\
 & {\scriptstyle f}\searrow \quad \swarrow{\scriptstyle j} & \\
 & X &
\end{array}
$$

and h *induces an isomorphism of* Y *onto an open subset of* $X_F(x_0)$.

Proof. Clear from the preceding discussion.

Note the general functorial behavior of the integral manifold. If

$$g\colon X \to X'$$

is an isomorphism, and F is an integrable tangent subbundle over X, then $F' = (Tg)(F) = g_*F$ is an integrable bundle over X'. Then the following diagram is commutative.

$$
\begin{array}{ccc}
X_F & \xrightarrow{\ g_F\ } & X'_{F'} \\
{\scriptstyle j}\downarrow & & \downarrow{\scriptstyle j'} \\
X & \xrightarrow{\ g\ } & X'
\end{array}
$$

The map g_F is, of course, the map having the same values as g, but viewed as a map on the manifold X_F.

§5. *Lie groups and subgroups*

It is not our purpose here to delve extensively into Lie groups, but to lay the groundwork for their theory. For more results, we refer the reader to texts on Lie groups, differential geometry, and also to the paper by W. Graeub, "Liesche Gruppen und affin zusammenhangende Mannigfaltigkeiten," *Acta Mathematica*, 1961, pp. 65–111. Although seemingly written to apply only to the finite dimensional case, this paper holds essentially in its entirety for the Banach case (and Hilbert case when dealing with Riemannian metrics), and is written on foundations corresponding to those of the present book.

By a **group manifold**, or a **Lie group** G, we mean a manifold with a group structure, that is a law of composition and inverse,

$$\tau : G \times G \to G \qquad \text{and} \qquad G \to G$$

which are morphisms. Thus each $x \in G$ gives rise to a left translation

$$\tau^x : G \to G$$

such that $\tau^x(y) = xy$.

When dealing with groups, we shall have to distinguish between isomorphisms in the category of manifolds, and isomorphisms in the category of group manifolds, which are also group homomorphisms. Thus we shall use prefixes, and speak of group manifold isomorphism, or manifold isomorphism as the case may be. We abbreviate these by GM-isomorphism or M-isomorphism. We see that left translation is an M-isomorphism, but not a GM-isomorphism.

Let e denote the origin (unit element) of G. If $v \in T_e G$ is a tangent vector at the origin, then we can translate it, and we obtain a map

$$(x, v) \mapsto \tau^x_* v = \xi_v(x)$$

which is easily verified to be a VB-isomorphism

$$G \times T_e G \to TG$$

from the product bundle to the tangent bundle of G. This is done at once using charts. Recall that $T_e G$ can be viewed as a Banachable space, using any local trivialization of G at e to get a toplinear isomorphism of $T_e G$ with the standard Banachable space on which G is modelled. Thus we see that the tangent bundle of a Lie group is trivializable.

A vector field ξ over G is called **left invariant** if $\tau^x_* \xi = \xi$ for all $x \in G$. Note that the map

$$x \mapsto \xi_v(x)$$

described above is a left invariant vector field, and that the association

$$v \mapsto \xi_v$$

obviously establishes a linear isomorphism between T_eG and the vector space of left invariant vector fields on G. The space of such vector fields will be denoted by $\mathfrak{g}$ or $\mathfrak{l}(G)$, and will be called the **Lie algebra** of G, because of the following result.

Proposition 3. Let ξ, η be left invariant vector fields on G. Then $[\xi, \eta]$ is also left invariant.

Proof. This follows from the general functorial formula

$$\tau_*^x[\xi, \eta] = [\tau_*^x \xi, \tau_*^x \eta] = [\xi, \eta].$$

Under the linear isomorphism of T_eG with $\mathfrak{l}(G)$, we can view $\mathfrak{l}(G)$ as a Banachable space. By a **Lie subalgebra** of $\mathfrak{l}(G)$ we shall mean a closed subspace $\mathfrak{h}$ which splits, and having the property that if $\xi, \eta \in \mathfrak{h}$, then $[\xi, \eta] \in \mathfrak{h}$ also.

Note: In the finite dimensional case, every subspace is closed and splits, so that only this last condition about the bracket product need be mentioned explicitly.

Let G, H be Lie groups. A map

$$f : H \to G$$

will be called a **homomorphism** if it is a group homomorphism and a morphism in the category of manifolds. Such a homomorphism induces a continuous linear map

$$T_e f = f_* : T_e H \to T_e G,$$

and it is clear that it also induces a corresponding linear map

$$\mathfrak{l}(H) \to \mathfrak{l}(G),$$

also denoted by f_*. Namely, if $v \in T_e H$ and ξ_v is the left invariant vector field on H induced by v, then

$$f_* \xi_v = \xi_{f_* v}.$$

The general functorial property of related vector fields applies to this case, and shows that the induced map

$$f_* : \mathfrak{l}(H) \to \mathfrak{l}(G)$$

is also a Lie algebra homomorphism, namely for $\xi, \eta \in I(H)$ we have

$$f_*[\xi, \eta] = [f_*\xi, f_*\eta].$$

Now suppose that the homomorphism $f: H \to G$ is also an immersion at the origin of H. Then by translation, one sees that it is an immersion at every point. If in addition it is an injective immersion, then we shall say that f is a **Lie subgroup** of G. We see that in this case, f induces a splitting injection

$$f_*: I(H) \to I(G).$$

The image of $I(H)$ in $I(G)$ is a Lie subalgebra of $I(G)$.

In general, let $\mathfrak{h}$ be a Lie subalgebra of $I(G)$ and let F_e be the corresponding subspace of T_eG. For each $x \in G$, let

$$F_x = \tau_*^x F_e.$$

Then F_x is a split subspace of T_xG, and using local charts, it is clear that the collection $F = \{F_x\}$ is a subbundle of TG, which is left invariant. Furthermore, if

$$f: H \to G$$

is a homomorphism which is an injective immersion, and if $\mathfrak{h}$ is the image of $I(H)$, then we also see that f is an integral manifold for the subbundle F. We shall now see that the converse holds, using Frobenius' theorem.

Theorem 5. *Let G be a Lie group, $\mathfrak{h}$ a Lie subalgebra of $I(G)$, and let F be the corresponding left invariant subbundle of TG. Then F is integrable.*

Proof. I owe the proof to Alan Weinstein. It is based on the following lemma.

Lemma. *Let X be a manifold, let ξ, η be vector fields at a point x_0, and let F be a subbundle of TX. If $\xi(x_0) = 0$ and ξ is contained in F, then $[\xi, \eta](x_0) \in F$.*

Proof. We can deal with the local representations, such that $X = U$ is open in $\mathbf{E}$, and F corresponds to a factor, that is

$$TX = U \times \mathbf{F}_1 \times \mathbf{F}_2 \quad \text{and} \quad F = U \times \mathbf{F}_1.$$

We may also assume without loss of generality that $x_0 = 0$. Then $\xi(0) = 0$, and $\xi: U \to \mathbf{F}_1$ may be viewed as a map into $\mathbf{F}_1$. We may write

$$\xi(x) = A(x)x,$$

with a morphism $A: U \to L(\mathbf{E}, \mathbf{F}_1)$. Indeed,

$$\xi(x) = \int_0^1 \xi'(tx) \, dt \cdot x,$$

and $A(x) = pr_1 \circ \int_0^1 \xi'(tx) \, dt$, where pr_1 is the projection on $\mathbf{F}_1$. Then

$$[\xi, \eta](x) = \eta'(x)\xi(x) - \xi'(x)\eta(x)$$
$$= \eta'(x)A(x)x - A'(x) \cdot x \cdot \eta(x) - A(x) \cdot \eta(x),$$

whence

$$[\xi, \eta](0) = A(0)\eta(0).$$

Since $A(0)$ maps $\mathbf{E}$ into $\mathbf{F}_1$, we have proved our lemma.

Back to the proof of the proposition. Let ξ, η be vector fields at a point x_0 in G, both contained in the invariant subbundle F. There exist invariant vector fields ξ_0 and η_0 at x_0 such that

$$\xi(x_0) = \xi_0(x_0) \qquad \text{and} \qquad \eta(x_0) = \eta_0(x_0).$$

Let

$$\xi_1 = \xi - \xi_0 \qquad \text{and} \qquad \eta_1 = \eta - \eta_0.$$

Then ξ_1, η_1 vanish at x_0 and lie in F. We get:

$$[\xi, \eta] = \sum_{i,j} [\xi_i, \eta_j].$$

The proposition now follows at once from the lemma.

Theorem 6. *Let G be a Lie group, let $\mathfrak{h}$ be a Lie subalgebra of $\mathrm{I}(G)$, and let F be its associated invariant subbundle. Let*

$$j: H \to G$$

be the maximal connected integral manifold of F passing through e. Then H is a subgroup of G, and $j: H \to G$ is a Lie subgroup of G. The association between $\mathfrak{h}$ and $j: H \to G$ establishes a bijection between Lie subalgebras of $\mathrm{I}(G)$ and Lie subgroups of G.

Proof. Let $x \in H$. The M-isomorphism τ^x induces a VB-isomorphism of F onto itself, in other words, F is invariant under τ_*^x. Furthermore, since H passes through e, and xe lies in H, it follows that $j: H \to G$ is also the maximal connected integral manifold of F passing through x. Hence x maps H onto itself. From this we conclude that if $y \in H$, then $xy \in H$, and there exists some $y \in H$ such that $xy = e$, whence $x^{-1} \in H$. Hence H is a subgroup. The other assertions are then clear.

If H is a Lie subgroup of G, belonging to the Lie algebra $\mathfrak{h}$, and F is the associated integrable left invariant tangent subbundle, then the integral manifold for F passing through a given point x is simply the translation xH, as one sees from first functorial principles.

When $\mathfrak{g}$ is 1-dimensional, then it is easy to see that the Lie subgroup is in fact a homomorphic image of an integral curve

$$\alpha : \mathbf{R} \to G$$

which is a homomorphism, and such that $\alpha'(0) = v$ is any vector in $T_e G$ which is the value at e of a non-zero element of $\mathfrak{h}$. Changing this vector merely reparametrizes the curve. The integral curve may coincide with the subgroup, or it comes back on itself, and then the subgroup is essentially a circle. Thus the integral curve need not be equal to the subgroup. However, locally near $t = 0$, they do coincide. Such an integral curve is called a **one-parameter subgroup** of G.

Using Theorem 1 of Chapter V, §1, it is then easy to see that if the Lie algebra of a connected Lie group G is commutative, then G itself is commutative. One first proves this for elements in a neighborhood of the origin, using 1-parameter subgroups, and then one gets the statement globally by expressing G as a union of products

$$UU \cdots U,$$

where U is a symmetric connected open neighborhood of the unit element. All of these statements are easy to prove, and belong to the first chapter of a book on Lie groups. Our purpose here is merely to lay the general foundations essentially belonging to general manifold theory.

Warning. The group of differential automorphisms of a finite dimensional manifold is "infiinte dimensional" but usually not a Lie group, because multiplication is usually continuous only in each variable separately. For an analysis of this, also in the context of H^p (Sobolev) spaces, cf. Ebin and Marsden [7].

CHAPTER VII

Riemannian Metrics

In our discussion of vector bundles, we put no greater structure on the fibers than that of topological vector space (of the same category as those used to build up manifolds). One can strengthen the notion so as to include the metric structure, and we are thus led to consider Hilbert bundles, whose fibers are Hilbert spaces.

Aside from the definitions, and basic properties, we deal with two special topics. On the one hand, we complete our uniqueness theorem on tubular neighborhoods by showing that when a Riemannian metric is given, a tubular neighborhood can be straightened out to a metric one. Secondly, we show how a Riemannian metric gives rise in a natural way to a spray, and thus how one recovers goedesics. The fundamental 2-form is used to identify the vector fields and 1-forms on the tangent bundle, identified with the cotangent bundle by the Riemannian metric.

We assume throughout that our manifolds are Hausdorff and are sufficiently differentiable that all our statements make sense. (For instance, when dealing with sprays, we take $p \geq 3$.)

Of necessity, we shall use the standard spectral theorem for (bounded) symmetric operators. A self-contained treatment will be given in the appendix.

§1. Definition and functoriality

Let $\mathbf{E}$ be a Hilbertable vector space, that is a topological vector space which is complete, and whose topology can be defined by the norm associated with a bilinear form, which is symmetric and positive definite. All facts needed in the sequel concerning Hilbert spaces can be found in the Appendix.

We consider $L_s^2(\mathbf{E})$, the set of continuous bilinear forms

$$\lambda : \mathbf{E} \times \mathbf{E} \to \mathbf{R}$$

which are symmetric. If x is fixed in $\mathbf{E}$, then the continuous linear form $\lambda_x(y) = \lambda(x, y)$ is given by an element of $\mathbf{E}$ which we denote by Ax, where

A is a continuous linear map of $\mathbf{E}$ into itself. The symmetry of λ implies that A is symmetric, that is we have

$$\langle Ax, y \rangle = \langle x, Ay \rangle$$

for all $x, y \in \mathbf{E}$. Conversely, given a symmetric continuous linear map $A : \mathbf{E} \to \mathbf{E}$ we can define a continuous bilinear form on $\mathbf{E}$ by this formula. Thus $L_s^2(\mathbf{E})$ is in bijection with the set of such operators, and is itself a Banach space, the norm being the usual operator norm.

The subset of $L_s^2(\mathbf{E})$ consisting of those forms corresponding to symmetric positive definite operators (by definition such that $A \geqq \varepsilon I$ for some $\varepsilon > 0$) will be called the **Riemannian** of $\mathbf{E}$ and be denoted by Ri($\mathbf{E}$). Forms λ in Ri($\mathbf{E}$) are called positive definite. The associated operator A of such a form is invertible, because its spectrum does not contain 0 and the continuous function $1/t$ is invertible on the spectrum.

In view of the operations on vector bundles (Chapter III, §4) we can apply the functor L_s^2 to any bundle whose fibers are Hilbertable spaces. If $\pi : E \to X$ is such a bundle, then we can form $L_s^2(\pi)$. A **metric** on π is then defined to be a section of $L_s^2(\pi)$. We know that the fiber of $L_s^2(\pi)$ at x is simply $L_s^2(\pi_x)$, and hence this fiber contains Ri(π_x). A metric g will be called a **Riemannian metric** if $g(x)$ lies in Ri(π_x) for each x in X, in other words, if $g(x)$ is a bilinear, symmetric, positive definite form on π_x.

Observe that the sections of $L_s^2(\pi)$ form a vector space (abstract) but that the Riemannian metrics do not. They form a convex cone. Indeed, if $a, b > 0$ and g_1, g_2 are two Riemannian metrics, then $ag_1 + bg_2$ is also a Riemannian metric.

Suppose we are given a VB-trivialisation of π over an open subset U of X, say

$$\tau : \pi^{-1}(U) \to U \times \mathbf{E}.$$

We can transport a given Riemannian metric g (or rather its restriction to $\pi^{-1}(U)$) to $U \times \mathbf{E}$. In the local representation, this means that for each $x \in U$ we can identify $g(x)$ with a symmetric positive definite operator A_x giving rise to the metric. Furthermore, the map

$$x \mapsto A_x$$

from U into the Banach space $L(\mathbf{E}, \mathbf{E})$ is a morphism.

As a matter of notation, we sometimes write g_x instead of $g(x)$. Thus if v, w are two vectors in E_x, then $g_x(v, w)$ is a number, and is more convenient to write than $g(x)(v, w)$. We shall also write $\langle v, w \rangle_x$ if the metric g is fixed once for all.

Proposition 1. *Let X be a manifold admitting partitions of unity. Let $\pi \colon E \to X$ be a vector bundle whose fibers are Hilbertable vector spaces. Then π admits a Riemannian metric.*

Proof. Find a partition of unity $\{U_i, \varphi_i\}$ such that $\pi \mid U_i$ is trivial, that is such that we have a trivialisation

$$\tau_i \colon \pi^{-1}(U_i) \to U_i \times \mathbf{E}$$

(working over a connected component of X, so that we may assume the fibers toplinearly isomorphic to a fixed Hilbert space $\mathbf{E}$). We can then find a Riemannian metric on $U_i \times \mathbf{E}$ in a trivial way. By transport of structure, there exists a Riemannian metric g_i on $\pi \mid U_i$ and we let $g = \sum \varphi_i g_i$. Then g is a Riemannian metric on π.

Let us investigate the functorial behavior of metrics.

Consider a VB-morphism

$$\begin{array}{ccc} E' & \xrightarrow{\ f\ } & E \\ {\scriptstyle \pi'}\downarrow & & \downarrow{\scriptstyle \pi} \\ X & \xrightarrow[\ f_0\ ]{} & Y \end{array}$$

with vector bundles E' and E over X and Y respectively, whose fibers are Hilbertable spaces. Let g be a metric on π, so that for each $y \in Y$ we have a continuous, bilinear, symmetric map

$$g(y) \colon E_y \times E_y \to \mathbf{R}.$$

Then the composite map

$$E'_x \times E'_x \to E_y \times E_y \to \mathbf{R}$$

with $y = f(x)$ is a metric on E'_x and one verifies immediately that it gives rise to a metric on the vector bundle π', which will be denoted by $f^*(g)$. The vector space of metrics on π will be denoted by $\mathrm{Met}(\pi)$ and the Riemannian metrics on π will be denoted by $\mathrm{Ri}(\pi)$. Then f induces a map

$$\mathrm{Met}(f) = f^* \colon \mathrm{Met}(\pi) \to \mathrm{Met}(\pi').$$

Furthermore, if f_x is injective and splits for each x in X, and g is a Riemannian metric, then obviously so is $f^*(g)$, and we can view f^* as mapping $\mathrm{Ri}(\pi)$ into $\mathrm{Ri}(\pi')$.

Let X be a manifold modelled on a Hilbertable space and let $T(X)$ be its tangent bundle. By abuse of language, we call a metric on $T(X)$ also a metric on X and write $\mathrm{Met}(X)$ instead of $\mathrm{Met}\big(T(X)\big)$. Similarly, we write $\mathrm{Ri}(X)$ instead of $\mathrm{Ri}\big(T(X)\big)$

Let $f: X \to Y$ be an immersion. Then for each $x \in X$, the linear map

$$T_x f: T_x(X) \to T_{f(x)}(Y)$$

is injective, and splits, and thus we obtain a contravariant map

$$f^*: \mathrm{Ri}(Y) \to \mathrm{Ri}(X),$$

each Riemannian metric on Y inducing a Riemannian metric on X (and of course each metric on Y induces a metric on X).

§2. The Hilbert group

Let $\mathbf{E}$ be a Hilbert space. The group of toplinear automorphisms $\mathrm{Laut}(\mathbf{E})$ contains the group $\mathrm{Hilb}(\mathbf{E})$ of Hilbert automorphisms, that is those toplinear automorphisms which preserve the inner product:

$$\langle Av, Aw \rangle = \langle v, w \rangle$$

for all $v, w \in \mathbf{E}$. We note that A is Hilbertian if and only if $A^* A = I$.

As usual, we say that a linear continuous map $A: \mathbf{E} \to \mathbf{E}$ is **symmetric** if $A^* = A$ and that it is **skew symmetric** if $A^* = -A$. We have a direct sum decomposition of the Banach space $L(\mathbf{E}, \mathbf{E})$ in terms of the two closed subspaces of symmetric and skew-symmetric operators:

$$A = \tfrac{1}{2}(A + A^*) + \tfrac{1}{2}(A - A^*).$$

We denote by $\mathrm{Sym}(\mathbf{E})$ and $\mathrm{Sk}(\mathbf{E})$ the Banach spaces of symmetric and skew-symmetric maps respectively. The word **operator** will always mean continuous linear map of $\mathbf{E}$ into itself.

Proposition 2. *For all operators A, the series*

$$\exp(A) = I + A + \frac{A^2}{2!} + \cdots$$

converges. If A commutes with B, then

$$\exp(A + B) = \exp(A) \exp(B).$$

For all operators sufficiently close to the identity I, the series

$$\log(A) = \frac{(A - I)}{1} + \frac{(A - I)^2}{2} + \cdots$$

converges, and if A commutes with B, then

$$\log(AB) = \log(A) + \log(B).$$

Proof. Standard.

We leave it as an exercise to the reader to show that the exponential function gives a C^∞-morphism of $L(\mathbf{E}, \mathbf{E})$ into itself. Similarly, a function admitting a development in power series say around 0 can be applied to the set of operators whose bound is smaller than the radius of convergence of the series, and gives a C^∞-morphism.

Proposition 3. *If A is symmetric (resp. skew-symmetric), then* $\exp(A)$ *is symmetric positive definite (resp. Hilbertian). If A is toplinear automorphism sufficiently close to I and is positive definite symmetric (resp. Hilbertian), then* $\log(A)$ *is symmetric (resp. skew-symmetric).*

Proof. The proofs are straightforward. As an example, let us carry out the proof of the last statement. Suppose A is Hilbertian and sufficiently close to I. Then $A^*A = I$ and $A^* = A^{-1}$. Then

$$\log(A)^* = \frac{(A^* - I)}{1} + \cdots$$

$$= \log(A^{-1}).$$

If A is close to I, so is A^{-1}, so that these statements make sense. We now conclude by noting that $\log(A^{-1}) = -\log(A)$. All the other proofs are carried out in a similar fashion, taking a star operator in a series term by term, under conditions which insure convergence.

The exponential and logarithm functions give inverse C^∞ mappings between neighborhoods of 0 in $L(\mathbf{E}, \mathbf{E})$ and neighborhoods of I in $\mathrm{Laut}(\mathbf{E})$. Furthermore, the direct sum decomposition of $L(\mathbf{E}, \mathbf{E})$ into symmetric and skew-symmetric subspaces is reflected locally in a neighborhood of I by a C^∞ direct product decomposition into positive definite and Hilbertian automorphisms. This direct product decomposition can be translated multiplicatively to any toplinear automorphism, because if $A \in \mathrm{Laut}(\mathbf{E})$ and B is close to A, then

$$B = AA^{-1}B = A\left(I - (I - A^{-1}B)\right)$$

and $(I - A^{-1}B)$ is small. This proves:

Proposition 4. *The Hilbert group of automorphisms of* $\mathbf{E}$ *is a closed submanifold of* $\mathrm{Laut}(E)$.

In addition to this local result, we get a global one also:

Proposition 5. *The exponential map gives a C^∞-isomorphism from the space* Sym(**E**) *of symmetric endomorphisms of* **E** *and the space* Pos(**E**) *of symmetric positive definite automorphisms of* **E**.

Proof. We must construct its inverse, and for this we use the spectral theorem. Given A, symmetric positive definite, the analytic function $\log t$ is defined on the spectrum of A, and thus $\log A$ is symmetric. One verifies immediately that it is the inverse of the exponential function (which can be viewed in the same way). We can expand $\log t$ around a large positive number c, in a power series uniformly and absolutely convergent in an interval $0 < \varepsilon \leqq t \leqq 2c - \varepsilon$, to achieve our purposes.

Proposition 6. *The manifold of toplinear automorphisms of the Hilbert space* **E** *is C^∞-isomorphic to the product of the Hilbert automorphisms and the positive definite symmetric automorphisms, under the mapping*

$$\mathrm{Hilb}(\mathbf{E}) \times \mathrm{Pos}(\mathbf{E}) \to \mathrm{Laut}(\mathbf{E})$$

given by

$$(H, P) \to HP.$$

Proof. Our map is induced by a continuous bilinear map of

$$L(\mathbf{E}, \mathbf{E}) \times L(\mathbf{E}, \mathbf{E})$$

into $L(\mathbf{E}, \mathbf{E})$ and so is C^∞. We must construct an inverse, or in other words express any given toplinear automorphism A in a unique way as a product $A = HP$ where H is Hilbertian, P is symmetric positive definite, and both H, P depend C^∞ on A. This is done as follows. First we note that A^*A is symmetric positive definite (because $\langle A^*Av, v \rangle = \langle Av, Av \rangle$, and furthermore, A^*A is a toplinear automorphism, so that 0 cannot be in its spectrum, and hence $A^*A \geqq \varepsilon I > O$ since the spectrum is closed). We let

$$P = (A^*A)^{1/2}$$

and let $H = AP^{-1}$. Then H is Hilbertian, because

$$H^*H = (P^{-1})^*A^*AP^{-1} = I.$$

Both P and H depend differentiably on A since all constructions involved are differentiable.

There remains to be shown that the expression as a product is unique. If $A = H_1P_1$ where H_1, P_1 are Hilbertian and symmetric positive definite respectively, then

$$H^{-1}H_1 = PP_1^{-1},$$

and we get $H_2 = PP_1^{-1}$ for some Hilbertian automorphism H_2. By definition,

$$I = H_2^* H_2 = (PP_1^{-1})^* PP_1^{-1}$$

and from the fact that $P^* = P$ and $P_1^* = P_1$, we find

$$P^2 = P_1^2.$$

Taking the log, we find $2 \log P = 2 \log P_1$. We now divide by 2 and take the exponential, thus giving $P = P_1$ and finally $H = H_1$. This proves our proposition.

§3. Reduction to the Hilbert group

We define a new category of bundles, namely the **Hilbert bundles** over X, denoted by HB(X). As before, we would denote by HB(X, **E**) or HB(X, $\mathfrak{A}$) those Hilbert bundles whose fiber is a Hilbert space **E** or lies in a category $\mathfrak{A}$.

Let $\pi \colon E \to X$ be a vector bundle over X, and assume that it has a trivialisation $\{(U_i, \tau_i)\}$ with trivialising maps

$$\tau_i \colon \pi^{-1}(U_i) \to U_i \times \mathbf{E}$$

where **E** is a Hilbert space, such that each toplinear automorphism $(\tau_j \tau_i^{-1})_x$ is a Hilbert automorphism. Equivalently, we could also say that τ_{ix} is a Hilbert isomorphism. Such a trivialisation will be called a **Hilbert trivialisation**. Two such trivialisations are called **Hilbert-compatible** if their union is again a Hilbert trivialisation. An equivalence class of such compatible trivialisations constitutes what we call a **Hilbert bundle** over X. Any such Hilbert bundle determines a unique vector bundle, simply by taking the VB-equivalence class determined by the trivialisation.

Given a Hilbert trivialisation $\{(U_i, \tau_i)\}$ of a vector bundle π over X, we can define on each fiber π_x a Hilbert space structure. Indeed, for each x we select an open set U_i in which x lies, and then transport to π_x the scalar product in **E** by means of τ_{ix}. By assumption, this is independent of the choice of U_i in which x lies. Thus in a Hilbert bundle, we can assume that the fibers are Hilbert spaces, not only Hilbertable.

It is perfectly possible that several distinct Hilbert bundles determine the same vector bundle.

Any Hilbert bundle determining a given vector bundle π will be said to be a **reduction of π to the Hilbert group**.

We can make Hilbert bundles into a category, if we take for the HB-morphisms the VB-morphisms which are injective and split at each point, and which preserve the metric, again at each point.

Each reduction of a vector bundle to the Hilbert group determines a Riemannian metric on the bundle. Indeed, defining for each $x \in X$ and $v, w \in \pi_x$ the scalar product

$$g_x(v, w) = \langle \tau_{ix}v, \tau_{ix}w \rangle$$

with any Hilbert-trivialising map τ_{ix} such that $x \in U_i$, we get a morphism

$$x \mapsto g_x$$

of X into the sections of $L_s^2(\pi)$ which are positive definite. We also have the converse.

Theorem 1. *Let π be a vector bundle over a manifold X, and assume that the fibers of π are all toplinearly isomorphic to a Hilbert space $\mathbf{E}$. Then the above map, from reductions of π to the Hilbert group, into the Riemannian metrics, is a bijection.*

Proof. Suppose that we are given an ordinary VB-trivialisation $\{(U_i, \tau_i)\}$ of π. We must construct an HB-trivialisation. For each i, let g_i be the Riemannian metric on $U_i \times \mathbf{E}$ transported from $\pi^{-1}(U_i)$ by means of τ_i. Then for each $x \in U_i$, we have a positive definite symmetric operator A_{ix} such that

$$g_{ix}(v, w) = \langle A_{ix}v, w \rangle$$

for all $v, w \in \mathbf{E}$. Let B_{ix} be the square root of A_{ix}. We define the trivialisation σ_i by the formula

$$\sigma_{ix} = B_{ix}\tau_{ix}$$

and contend that $\{(U_i, \sigma_i)\}$ is a Hilbert trivialisation. Indeed, from the definition of g_{ix}, it suffices to verify that the VB-isomorphism

$$B_i: U_i \times \mathbf{E} \to U_i \times \mathbf{E}$$

given by B_{ix} on each fiber, carries g_i on the usual metric. But we have, for $v, w \in E$:

$$\langle B_{ix}v, B_{ix}w \rangle = \langle A_{ix}v, w \rangle$$

since B_{ix} is symmetric, and equal to the square root of A_{ix}. This proves what we want.

At this point, it is convenient to make an additional comment on normal bundles.

Let α, β be two Hilbert bundles over the manifold X, and let $f\colon \alpha \to \beta$ be an HB-morphism. Assume that

$$0 \to \alpha \xrightarrow{\ f\ } \beta$$

is exact. Then by using the Riemannian metric, there is a natural way of constructing a splitting for this sequence (cf. Chapter III, §5).

Using Theorem 1 of the Appendix, we see at once that if $\mathbf{F}$ is a (closed) subspace of a Hilbert space, then $\mathbf{E}$ is the direct sum

$$\mathbf{E} = \mathbf{F} \oplus \mathbf{F}^{\perp}$$

of $\mathbf{F}$ and its orthogonal complement, consisting of all vectors perpendicular to $\mathbf{F}$.

In our exact sequence, we may view f as an injection. For each x we let $\alpha_x^{\perp}$ be the orthogonal complement of α_x in β_x. Then we shall find an exact sequence of VB-morphisms

$$\beta \xrightarrow{\ h\ } \alpha \to 0$$

whose kernel is $\alpha^{\perp}$ (set theoretically). In this manner, the collection of orthogonal complements $\alpha_x^{\perp}$ can be given the structure of a Hilbert bundle.

For each x we can write $\beta_x = \alpha_x \oplus \alpha_x^{\perp}$ and we define h_x to be the projection in this direct sum decomposition. This gives us a mapping $h\colon \beta \to \alpha$, and it will suffice to prove that h is a VB-morphism. In order to do this, we may work locally. In that case, after taking suitable VB-automorphisms over a small open set U of X, we can assume that we deal with the following situation.

Our vector bundle β is equal to $U \times \mathbf{E}$ and α is equal to $U \times \mathbf{F}$ for some subspace $\mathbf{F}$ of $\mathbf{E}$, so that we can write $\mathbf{E} = \mathbf{F} \times \mathbf{F}^{\perp}$. Our HB-morphism is then represented for each x by an injection $f_x\colon \mathbf{F} \to \mathbf{E}$:

$$U \times \mathbf{F} \xrightarrow{\ f\ } U \times \mathbf{E}.$$

By the definition of exact sequences, we can find two VB-isomorphisms τ and σ such that the following diagram is commutative:

$$
\begin{array}{ccc}
U \times \mathbf{F} & \xrightarrow{\ f\ } & U \times \mathbf{E} \\
{\scriptstyle \sigma} \downarrow & & \downarrow {\scriptstyle \tau} \\
U \times \mathbf{F} & \longrightarrow & U \times \mathbf{E}
\end{array}
$$

and such that the bottom map is simply given by the ordinary inclusion of $\mathbf{F}$ in $\mathbf{E}$. We can transport the Riemannian structure of the bundles on top to the bundles on the bottom by means of σ^{-1} and τ^{-1} respectively. We are therefore reduced to the situation where f is given by the simple inclusion, and the Riemannian metric on $U \times \mathbf{E}$ is given by a family A_x of symmetric positive definite operators on $\mathbf{E}$ ($x \in U$). At each point x, we have $\langle v, w \rangle_x = \langle A_x v, w \rangle$. We observe that the map

$$A : U \times \mathbf{E} \to U \times \mathbf{E}$$

given by A_x on each fiber is a VB-automorphism of $U \times \mathbf{E}$. Let $\mathrm{pr}_\mathbf{F}$ be the projection of $U \times \mathbf{E}$ on $U \times \mathbf{F}$. It is a VB-morphism. Then the composite

$$h = \mathrm{pr}_\mathbf{F} \circ A$$

gives us a VB-morphism of $U \times \mathbf{E}$ on $U \times \mathbf{F}$, and the sequence

$$U \times \mathbf{E} \xrightarrow{h} U \times \mathbf{F} \to 0$$

is exact. Finally, we note that the kernel of h consists precisely of the orthogonal complement of $U \times \mathbf{F}$ in each fiber. This proves what we wanted.

§4. Hilbertian tubular neighborhoods

Let $\mathbf{E}$ be a Hilbert space. Then the open ball of radius 1 is isomorphic to $\mathbf{E}$ itself under the mapping

$$v \mapsto \frac{v}{(1 - |v|^2)^{1/2}},$$

the inverse mapping being

$$w \mapsto \frac{w}{(1 + |w|^2)^{1/2}}.$$

If $a > 0$, then any ball of radius a is isomorphic to the unit ball under multiplication by the scalar a (or a^{-1}).

Let X be a manifold, and $\sigma : X \to \mathbf{R}$ a function (morphism) such that $\sigma(x) > 0$ for all $x \in X$. Let $\pi : E \to X$ be a Hilbert bundle over X. We denote by $E(\sigma)$ the subset of E consisting of those vectors v such that, if v lies in E_x, then

$$|v|_x < \sigma(x).$$

Then $E(\sigma)$ is an open neighborhood of the zero section.

Proposition 7. *Let X be a manifold and $\pi: E \to X$ a Hilbert bundle. Let $\sigma: X \to \mathbf{R}$ be a morphism such that $\sigma(x) > 0$ for all x. Then the mapping*

$$w \to \frac{\sigma(\pi w)w}{(1 + |w|^2)^{1/2}}$$

gives an isomorphism of E onto $E(\sigma)$.

Proof. Obvious. The inverse mapping is constructed in the obvious way.

Corollary. *Let X be a manifold admitting partitions of unity, and let $\pi: E \to X$ be a Hilbert bundle over X. Then E is compressible.*

Proof. Let Z be an open neighborhood of the zero section. For each $x \in X$, there exists an open neighborhood V_x and a number $a_x > 0$ such that the vectors in $\pi^{-1}(V_x)$ which are of length $< a_x$ lie in Z. We can find a partition of unity $\{(U_i, \varphi_i)\}$ on X such that each U_i is contained in some $V_{x(i)}$. We let σ be the function

$$\sum a_{x(i)}\varphi_i.$$

Then $E(\sigma)$ is contained in Z, and our assertion follows from the proposition.

Proposition 8. *Let X be a manifold. Let $\pi: E \to X$ and $\pi_1: E_1 \to X$ be two Hilbert bundles over X. Let*

$$\lambda: E \to E_1$$

be a VB-isomorphism. Then there exists an isotopy of VB-isomorphisms

$$\lambda_t: E \to E_1$$

with proper domain $[0, 1]$ such that $\lambda_1 = \lambda$ and λ_0 is an HB-isomorphism.

Proof. We find reductions of E and E_1 to the Hilbert group, with Hilbert trivialisations $\{(U_i, \tau_i)\}$ for E and $\{(U_i, \rho_i)\}$ for E_1. We can then factor $\rho_i \lambda \tau_i^{-1}$ as in Proposition 6 of §2, applied to each fiber map,

$$
\begin{array}{ccccc}
U_i \times \mathbf{E} & \longrightarrow & U_i \times \mathbf{E} & \longrightarrow & U_i \times \mathbf{E} \\
{\scriptstyle \tau_i}\big\uparrow & & {\scriptstyle \tau_i}\big\uparrow & & \big\uparrow{\scriptstyle \rho_i} \\
\pi^{-1}(U_i) & \xrightarrow{\ \lambda_P\ } & \pi(U_i^{-1}) & \xrightarrow{\ \lambda_H\ } & \pi_1^{-1}(U_i)
\end{array}
$$

and obtain a factorization of λ into $\lambda = \lambda_H \lambda_P$ where λ_H is a HB-isomorphism and λ_P is a positive definite symmetric VB-automorphism. The latter form a convex set, and our isotopy is simply

$$\lambda_t = \lambda_H \circ (tI + (1 - t)\lambda_P).$$

(Smooth out the end points if you wish.)

Theorem 2. *Let X be a submanifold of Y. Let $\pi\colon E \to X$ and $\pi_1\colon E_1 \to X$ be two Hilbert bundles. Assume that E is compressible. Let $f\colon E \to Y$ and $g\colon E_1 \to Y$ be two tubular neighborhoods of X in Y. Then there exists an isotopy*

$$f_t\colon E \to Y$$

of tubular neighborhoods with proper domain $[0, 1]$ and there exists an HB-isomorphism $\mu\colon E \to E_1$ such that $f_1 = f$ and $f_0 = g\mu$.

Proof. From Theorem 10 of Chapter IV, §6 we know already that there exists a VB-isomorphism λ such that $f \approx g\lambda$. Using the preceding proposition, we know that $\lambda \approx \mu$ where μ is a HB-isomorphism. Thus $g\lambda \approx g\mu$ and by transitivity, $f \approx g\mu$, as was to be shown.

Remark. In view of Proposition 7, we could of course replace the condition that E be compressible by the more useful condition (in practice) that X admit partitions of unity.

§5. *Non-singular bilinear tensors*

Let $\mathbf{E}$ be a Hilbert space, and

$$\Omega\colon \mathbf{E} \times \mathbf{E} \to \mathbf{R}$$

a continuous bilinear map. There exists a unique operator A such that

$$\Omega(v, w) = \langle Av, w \rangle$$

for all $v, w \in \mathbf{E}$. If A is invertible (i.e. there exists an operator A^{-1} such that $AA^{-1} = A^{-1}A = I$), then we shall say that Ω is **non-singular**.

Let Y be a manifold, and $\pi\colon E \to Y$ a Hilbert bundle over Y. Let Ω be a tensor field of type L^2 on E, that is to say a section of the bundle $L^2(E)$ (or $L^2(\pi)$). Then for each $y \in Y$, we have a bilinear continuous map Ω_y on E_y. If Ω_y is non-singular for each $y \in Y$, then we say that Ω is **non-singular**. If π is trivial, and we have a trivialisation $Y \times \mathbf{E}$, then the local representation of Ω can be described by a morphism of Y into the Banach space of operators. If Ω is non-singular, then the image of this morphism is contained in the open set of invertible operators. (If Ω is a 2-form, this image is contained in the submanifold of skew-symmetric operators.)

Let Ω be a non-singular tensor field of type L^2 on E, or as we shall also say, a non-singular **bilinear tensor field** on E. Then Ω can be used to identify the sections $\Gamma(E)$ of E and the 1-forms on E in the following manner. Let ξ be a section of E. For each $y \in Y$, we define a continuous linear map

$$(\Omega \circ \xi)_y : E_y \to \mathbf{R}$$

by the formula

$$\langle (\Omega \circ \xi)_y, w \rangle = \langle \Omega(y), \xi(y) \times w \rangle = \Omega_y\big(\xi(y), w\big).$$

Looking at local trivialisations of π, we see at once that $\Omega \circ \xi$ is a 1-form on E.

Conversely, let ω be a given 1-form on E. For each $y \in Y$, ω_y is therefore a 1-form on E_y and since Ω is non-singular, there exists a unique element $\xi(y)$ of E_y such that

$$\Omega_y\big(\xi(y), w\big) = \omega_y(w)$$

for all $w \in E_y$. In this fashion, we obtain a mapping ξ of Y into E and we contend that ξ is a morphism (and therefore a section).

To prove our contention we can look at the local representations. We use Ω and ω to denote these. They are represented over a suitable open set U by two morphisms

$$A : U \to \mathrm{Aut}(\mathbf{E}) \qquad \text{and} \qquad \eta : U \to \mathbf{E}$$

such that

$$\Omega_y(v, w) = \langle A_y v, w \rangle \qquad \text{and} \qquad \omega_y(w) = \langle \eta(y), w \rangle.$$

From this we see that

$$\xi(y) = A_y^{-1}\eta(y),$$

from which it is clear that ξ is a morphism. We may summarize our discussion as follows.

Proposition 9. *Let Y be a manifold and $\pi : E \to Y$ a Hilbert bundle over Y. Let Ω be a non-singular bilinear tensor field on E. Then Ω induces a bijection between sections of E and 1-forms on E. A section ξ corresponds to a 1-form ω if and only if*

$$\Omega \circ \xi = \omega.$$

We observe that our bijection is clearly a linear isomorphism, both the sections of E and the 1-forms being vector spaces over the constant field.

In many applications, one takes the differential form to be df for some function f. The vector field corresponding to df is then called the **gradient** of f. In the next section we shall work on the tangent bundle rather than the base manifold and in this manner get the vector field giving rise to geodesics.

§6. Riemannian metrics and sprays

Let X be a Riemannian manifold, modelled on the Hilbert space $\mathbf{E}$. The scalar product $\langle\,,\rangle$ in $\mathbf{E}$ identifies $\mathbf{E}$ with its dual $\mathbf{E}^*$. The Riemannian metric on X gives a toplinear isomorphism of each tangent space $T_x(X)$ with $T_x^*(X)$. If we work locally with $X = U$ open in $\mathbf{E}$ and we make the identification

$$T(U) = U \times \mathbf{E} \quad \text{and} \quad T^*(U) = U \times \mathbf{E}^*$$

then the metric gives a VB-isomorphism

$$\varphi : T(U) \to T^*(U)$$

by means of a morphism

$$g : U \to L(\mathbf{E}, \mathbf{E}^*)$$

such that $\varphi(x, v) = (x, g(x)v)$. For each $x \in U$ the scalar product given by the metric is then denoted by

$$\langle y, w \rangle_x = \langle y, g(x)w \rangle, \qquad y, w \in \mathbf{E}.$$

For each $x \in U$ we note that $Dg(x)$ maps $\mathbf{E}$ into $L(\mathbf{E}, \mathbf{E}^*)$, and for $x \in U$, $y \in \mathbf{E}$ we write

$$Dg(x)y \cdot v = Dg_{x,y}(v).$$

If we pull back the canonical 2-form described in Chapter V from $T^*(U)$ to $T(U)$ by means of φ, then its description locally can be written on $U \times \mathbf{E}$ in the following manner.

$$\langle \Omega_{(x,v)}, (y_1, w_1) \times (y_2, w_2) \rangle$$

$$= \langle y_2, w_1 \rangle_x - \langle y_1, w_2 \rangle_x + \langle Dg_{x,y_1}(v), y_2 \rangle - \langle Dg_{x,y_2}(v), y_1 \rangle.$$

because in our present local representation, $T\varphi$ maps an element (x, v, y, w) of $T\big(T(U)\big)$ on the element

$$(x, g(x)v, y, Dg_{x,y}(v) + g(x)w)$$

in $T\big(T^*(U)\big)$.

From the simple formula giving our fundamental 2-form on the cotangent bundle in Chapter V, we see at once that it is nonsingular on $T^*(U)$. Since

φ is a VB-isomorphism, it follows that the pull-back of this 2-form to the tangent bundle is also non-singular.

We shall now apply the results of the preceding section. To do so, we construct a 1-form on $T(X)$. Indeed, we have a function

$$K : T(X) \to \mathbf{R}$$

given by $K(v) = \frac{1}{2}\langle v, v \rangle_x$ if v is in T_x. Then dK is a 1-form. By Proposition 9 of the preceding section, it corresponds to a vector field on $T(X)$, and we contend:

Theorem 3. *The vector field on $T(X)$ corresponding to $-dK$ under the fundamental 2-form is a spray over X (known as the geodesic spray).*

Proof. We work locally, and use the criterion of Proposition 5 of Chapter IV, §3. We take U open in E and have the double tangent bundle

$$(U \times \mathbf{E}) \times (\mathbf{E} \times \mathbf{E})$$
$$\downarrow$$
$$U \times \mathbf{E}$$
$$\downarrow$$
$$U$$

Our function K can be written

$$K(x, v) = \tfrac{1}{2}\langle v, v \rangle_x = \tfrac{1}{2}\langle v, g(x)v \rangle,$$

and dK at a point (x, v) is simply the ordinary derivative

$$DK(x, v) : \mathbf{E} \times \mathbf{E} \to \mathbf{R}.$$

The derivative DK is completely described by the two partial derivatives, and we have

$$DK(x, v) \cdot (y, w) = D_1 K(x, v) \cdot y + D_2 K(x, v) \cdot w.$$

From the definition of derivative, we find

$$D_2 K(x, v) \cdot w = \langle v, w \rangle_x.$$

We use the notation of Proposition 5, Chapter IV, §3. We can represent the vector field corresponding to $-dK$ under the canonical 2-form Ω by a morphism $f : U \times \mathbf{E} \to \mathbf{E} \times \mathbf{E}$, which we write in terms of its two components:

$$f(x, v) = (f_1(x, v), f_2(x, v)).$$

By the preceding section, we have for all $(x, v) \in U \times \mathbf{E}$ and $(y, w) \in \mathbf{E} \times \mathbf{E}$:

$$\Omega_{x,v}\langle (f_1(x, v), f_2(x, v)), (y, w) \rangle = -\langle DK(x, v), (y, w) \rangle$$
$$= -D_1 K(x, v) \cdot y - \langle v, w \rangle_x.$$

In the expression obtained above for $\Omega_{x,v}$ we find that as a function of $w_2 = w$ it has only one term, namely

$$-\langle y_1, w_2 \rangle_x.$$

From this it follows that

$$-\langle f_1(x, v), w \rangle_x = -\langle v, w \rangle_x$$

for all w and hence that $f_1(x, v) = v$, whence our vector field is a second order differential equation.

We can now write down the second factor of our vector field again using the expression for $\Omega_{x,v}$ obtained above, and we must have

$$\langle y, f_2(x, v) \rangle_x = \langle Dg_{x,y}(v), v \rangle - \langle Dg_{x,v}(v), y \rangle - D_1 K(x, v) \cdot y.$$

From this one sees that f_2 is homogeneous of degree 2 in its second variable v, in other words that it represents a spray. This concludes the proof.

In terms of local coordinates, the Riemannian spray is given by a map f_2 satisfying the second order differential equation

$$\frac{d^2 x_i}{dt^2} = f_2(x, y) \qquad \text{and} \qquad y_i = \frac{dx_i}{dt}.$$

As a function of the variables y, the map f is quadratic, and its coefficients are functions of x, called the **Christoffel symbols**, Γ^i_{jk}. Thus by definition, the above differential equation is of type

$$\frac{d^2 x_i}{dt^2} = -\sum_{j,k} \Gamma^i_{jk}(x) \frac{dx_k}{dt} \frac{dx_j}{dt}.$$

In terms of the standard basis for $\mathbf{R}^n$, the Riemannian metric is then given by a matrix

$$(g_{ij}(x)),$$

and we let (g^{ij}) be the inverse matrix. Then the last formula relating the Riemannian metric and dK in the above proof can be written in terms of the local coordinates in terms of the Christoffel symbols, namely

$$-\Gamma^j_{ik} = \frac{1}{2} \sum_v g^{jv} \left(\frac{\partial g_{ik}}{\partial x_v} - \frac{\partial g_{vk}}{\partial x_i} - \frac{\partial g_{iv}}{\partial x_k} \right).$$

§7. The Morse-Palais lemma

Let U be an open set in some (real) Hilbert space $\mathbf{E}$, and let f be a C^{p+2} function on U, with $p \geq 1$. We say that x_0 is a **critical point** for f if $Df(x_0) = 0$. We wish to investigate the behavior of f at a critical point. After translations, we can assume that $x_0 = 0$ and that $f(x_0) = 0$. We observe that the second derivative $D^2 f(0)$ is a continuous bilinear form on $\mathbf{E}$. Let $\lambda = D^2 f(0)$, and for each $x \in \mathbf{E}$ let λ_x be the functional $y \mapsto \lambda(x, y)$. If the map $x \mapsto \lambda_x$ is a toplinear isomorphism of $\mathbf{E}$ with its dual space $\mathbf{E}^*$, then we say that λ is **non-singular**, and we say that the critical point is **non-degenerate**.

We recall that a local C^p-isomorphism φ at 0 is a C^p-invertible map defined on an open set containing 0.

Theorem 4. *Let f be a C^{p+2} function defined on an open neighborhood of 0 in the Hilbert space $\mathbf{E}$, with $p \geq 1$. Assume that $f(0) = 0$, and that 0 is a non-degenerate critical point of f. Then there exists a local C^p-isomorphism at 0, say φ, and an invertible symmetric operator A such that*

$$f(x) = \langle A\varphi(x), \varphi(x) \rangle.$$

Proof. We may assume that U is a ball around 0. We have

$$f(x) = f(x) - f(0) = \int_0^1 Df(tx)x\, dt,$$

and applying the same formula to Df instead of f, we get

$$f(x) = \int_0^1 \int_0^1 D^2 f(stx)tx \cdot x\, ds\, dt = g(x)(x, x)$$

where

$$g(x) = \int_0^1 \int_0^1 D^2 f(stx)t\, ds\, dt.$$

Then g is a C^p map into the Banach space of continuous bilinear maps on $\mathbf{E}$, and even the space of symmetric such maps. We know that this Banach space is toplinearly isomorphic to the space of symmetric operators on $\mathbf{E}$, and thus we can write

$$f(x) = \langle A(x)x, x \rangle$$

where $A : U \to \mathrm{Sym}\,(\mathbf{E})$ is a C^p map of U into the space of symmetric operators on $\mathbf{E}$. A straightforward computation shows that

$$D^2 f(0)(v, w) = \langle A(0)v, w \rangle.$$

Since we assumed that $D^2f(0)$ is non-singular, this means that $A(0)$ is invertible, and hence $A(x)$ is invertible for all x sufficiently near 0.

We want to define $\varphi(x)$ to be $C(x)x$ where C is a suitable C^p map from a neighborhood of 0 into the open set of invertible operators, and in such a way that we have

$$\langle A(x)x, x \rangle = \langle A(0)\varphi(x), \varphi(x) \rangle = \langle A(0)C(x)x, C(x)x \rangle.$$

This means that we must seek a map C such that

$$C(x)^* A(0) C(x) = A(x).$$

If we let $B(x) = A(0)^{-1}A(x)$, then $B(x)$ is close to the identity I for small x. The square root function has a power series expansion near 1, which is a uniform limit of polynomials, and is C^∞ on a neighborhood of I, and we can therefore take the square root of $B(x)$, so that we let

$$C(x) = B(x)^{1/2}.$$

We contend that this $C(x)$ does what we want. Indeed, since both $A(0)$ and $A(x)$ $\left(\text{or } A(x)^{-1}\right)$ are self-adjoint, we find that

$$B(x)^* = A(x)A(0)^{-1},$$

whence

$$B(x)^* A(0) = A(0)B(x).$$

But $C(x)$ is a power series in $I - B(x)$, and $C(x)^*$ is the same power series in $I - B(x)^*$. The preceding relation holds if we replace $B(x)$ by any power of $B(x)$ (by induction), hence it holds if we replace $B(x)$ by any polynomial in $I - B(x)$, and hence finally, it holds if we replace $B(x)$ by $C(x)$, and thus

$$C(x)^* A(0) C(x) = A(0)C(x)C(x) = A(0)B(x) = A(x),$$

which is the desired relation.

All that remains to be shown is that φ is a local C^p-isomorphism at 0. But one verifies that in fact, $D\varphi(0) = C(0)$, so that what we need follows from the inverse mapping theorem. This concludes the proof of Theorem 4.

Corollary. *Let f be a C^{p+2} function near 0 on the Hilbert space* **E**, *such that 0 is a non-degenerate critical point. Then there exists a local C^p-isomorphism ψ at 0, and an orthogonal decomposition* **E** $=$ **F** $+$ **F**$^\perp$, *such that if we write $\psi(x) = y + z$ with $y \in$ **F** and $z \in$ **F**$^\perp$, then*

$$f(\psi(x)) = \langle y, y \rangle - \langle z, z \rangle.$$

Proof. On a space where A is positive definite, we can always make the toplinear isomorphism $x \mapsto A^{1/2}x$ to get the quadratic form to become the given hermitian product $\langle , \rangle$, and similarly on a space where A is negative definite. In general, we use the spectral theorem to decompose $\mathbf{E}$ into a direct orthogonal sum such that the restriction of A to the factors is positive definite and negative definite respectively.

Note. The Morse-Palais lemma was proved originally by Morse in the finite dimensional case, using the Gram-Schmidt orthogonalization process. The elegant generalization and its proof in the Hilbert space case is due to Palais. It shows (in the language of coordinate systems) that a function near a critical point can be expressed as a quadratic form after a suitable change of coordinate system (satisfying requirements of differentiability). It comes up naturally in the calculus of variations. For instance, one considers a space of paths (of various smoothness) $\sigma: [a, b] \to E$ where E is a Hilbert space. One then defines a function on these paths, essentially related to the length,

$$f(\sigma) = \int_a^b \langle \sigma'(t), \sigma'(t) \rangle \, dt,$$

and one investigates the critical points of this function, especially its minimum values. These turn out to be the solutions of the variational problem, by definition of what one means by a variational problem. Even if E is finite dimensional, so a Euclidean space, the space of paths is infinite dimensional.

CHAPTER VIII

Integration of Differential Forms

Throughout this chapter, μ is Lebesgue measure on $\mathbf{R}^n$.
If A is a subset of $\mathbf{R}^n$, we write $\mathscr{L}^1(A)$ instead of $\mathscr{L}^1(A, \mu, \mathbf{C})$.
All manifolds are assumed finite dimensional.
They may have a boundary.

§1. Sets of measure 0

We recall that a set has measure 0 in $\mathbf{R}^n$ if and only if, given ε, there exists a covering of the set by a sequence of rectangles $\{R_j\}$ such that $\sum \mu(R_j) < \varepsilon$. We denote by R_j the closed rectangles, and we may always assume that the interiors R_j^0 cover the set, at the cost of increasing the lengths of the sides of our rectangles very slightly (an $\varepsilon/2^n$ argument). We shall prove here some criteria for a set to have measure 0. We leave it to the reader to verify that instead of rectangles, we could have used cubes in our characterization of a set of a measure 0 (a cube being a rectangle all of whose sides have the same length).

We recall that a map f satisfies a **Lipschitz condition** on a set A if there exists a number C such that

$$|f(x) - f(y)| \leq C|x - y|$$

for all $x, y \in A$. Any C^1 map f satisfies locally at each point a Lipschitz condition, because its derivative is bounded in a neighborhood of each point, and we can then use the mean value estimate,

$$|f(x) - f(y)| \leq |x - y| \sup |f'(z)|,$$

the sup being taken for z on the segment between x and y. We can take the neighborhood of the point to be a ball, say, so that the segment between any two points is contained in the neighborhood.

Lemma 1. *Let A have measure 0 in $\mathbf{R}^n$ and let $f : A \to \mathbf{R}^n$ satisfy a Lipschitz condition. Then $f(A)$ has measure 0.*

171

Proof. Let C be a Lipschitz constant for f. Let $\{R_j\}$ be a sequence of cubes covering A such that $\sum \mu(R_j) < \varepsilon$. Let r_j be the length of the side of R_j. Then for each j we see that $f(A \cap S_j)$ is contained in a cube R'_j whose sides have length $\leq 2Cr_j$. Hence

$$\mu(R'_j) \leq 2^n C^n r_j^n = 2^n C^n \mu(R_j).$$

Our lemma follows.

Lemma 2. *Let U be open in $\mathbf{R}^n$ and let $f: U \to \mathbf{R}^n$ be a C^1 map. Let Z be a set of measure 0 in U. Then $f(Z)$ has measure 0.*

Proof. For each $x \in U$ there exists a rectangle R_x contained in U such that the family $\{R_x^0\}$ of interiors covers Z. Since U is separable, there exists a denumerable subfamily covering Z, say $\{R_j\}$. It suffices to prove that $f(Z \cap R_j)$ has measure 0 for each j. But f satisfies a Lipschitz condition on R_j since R_j is compact and f' is bounded on R_j, being continuous. Our lemma follows from Lemma 1.

Lemma 3. *Let A be a subset of $\mathbf{R}^m$. Assume that $m < n$. Let $f: A \to \mathbf{R}^n$ satisfy a Lipschitz condition. Then $f(A)$ has measure 0.*

Proof. We view $\mathbf{R}^m$ as embedded in $\mathbf{R}^n$ on the space of the first m co-ordinates. Then $\mathbf{R}^m$ has measure 0 in $\mathbf{R}^n$, so that A has also n-dimensional measure 0. Lemma 3 is therefore a consequence of Lemma 1.

Note. All three lemmas may be viewed as stating that certain parametrized sets have measure 0. Lemma 3 shows that parametrizing a set by strictly lower dimensional spaces always yields an image having measure 0. The other two lemmas deal with a map from one space into another of the same dimension. Observe that Lemma 3 would be false if f is only assumed to be continuous (Peano curves).

The next theorem will be used later only in the proof of the residue theorem, but it is worthwhile inserting it at this point.

Let $f: X \to Y$ be a morphism of class C^p, with $p \geq 1$, and assume throughout this section that X, Y are finite dimensional. A point $x \in X$ is called a **critical point** of f if f is not submersion at x. This means that

$$T_x f: T_x X \to T_{f(x)} Y$$

is not surjective, according to our differential criterion for a submersion.

Assume that a manifold X has a countable base for its charts. Then we can say that a set has measure 0 in X if its intersection with each chart has measure 0.

Sard's theorem. *Let* $f: X \to Y$ *be a* C^∞ *morphism of finite dimensional manifolds having a countable base. Let* Z *be the set of critical points of* f *in* X. *Then* $f(Z)$ *has measure* 0 *in* Y.

Proof. (Due to Dieudonné.) By induction on the dimension n of X. The assertion is trivial if $n = 0$. Assume $n \geq 1$. It will suffice to prove the theorem locally in the neighborhood of a point in Z. We may assume that $X = U$ is open in $\mathbf{R}^n$ and

$$f: U \to \mathbf{R}^p$$

can be expressed in terms of coordinate functions,

$$f = (f_1, \ldots, f_p).$$

We let as usual

$$D^\alpha = D_1^{\alpha_1} \cdots D_n^{\alpha_n}$$

be a differential operator, and call $|\alpha| = \alpha_1 + \cdots + \alpha_n$ its **order**. We let $Z_0 = Z$ and for $m \geq 1$ we let Z_m be the set of points $x \in Z$ such that

$$D^\alpha f_j(x) = 0$$

for all $j = 1, \ldots, p$ and all α with $1 \leq |\alpha| \leq m$. We shall prove:

(1) *For each* $m \geq 0$ *the set* $f(Z_m - Z_{m+1})$ *has measure* 0.

(2) *If* $m \geq n/p$, *then* $f(Z_m)$ *has measure* 0.

This will obviously prove Sard's theorem.

Proof of 1. Let $a \in Z_m - Z_{m+1}$. Suppose first that $m = 0$. Then for some coordinate function, say $j = 1$, and after a renumbering of the variables if necessary, we have

$$D_1 f_1(a) \neq 0.$$

The map

$$g: x \mapsto \big(f_1(x), x_2, \ldots, x_p\big)$$

obviously has an invertible derivative at $x = a$, and hence is a local isomorphism at a. Considering $f \circ g^{-1}$ instead of f, we are reduced to the case where f is given by

$$f(x) = \big(x_1, f_2(x), \ldots, f_p(x)\big) = \big(x_1, h(x)\big),$$

where h is the projection of f on the last $p - 1$ coordinates and is therefore a morphism $h\colon V \to \mathbf{R}^{p-1}$ defined on some open V containing a. Then

$$Df(x) = \begin{pmatrix} 1 & 0 \\ * & Dh(x) \end{pmatrix}.$$

From this it is clear that x is a critical point for f if and only if x is a critical point for h, and it follows that $h(Z \cap V)$ has measure 0 in $\mathbf{R}^{p-1}$. Since $f(Z)$ is contained in $\mathbf{R}^1 \times h(Z)$, we conclude that $f(Z)$ has measure 0 in $\mathbf{R}^p$ as desired.

Next suppose that $m \geqq 1$. Then for some α with $|\alpha| = m + 1$, and say $j = 1$, we have

$$D^\alpha f_1(a) \neq 0.$$

Again after a renumbering of the indices, we may write

$$D^\alpha f_1 = D_1 g_1$$

for some function g_1, and we observe that $g_1(x) = 0$ for all $x \in Z_m$, in a neighborhood of a. The map

$$g\colon x \mapsto \bigl(g_1(x), x_2, \ldots, x_n\bigr)$$

is then a local isomorphism at a, say on an open set V containing a, and we see that

$$g(Z_m \cap V) \subset \{0\} \times R^{n-1}.$$

We view g as a change of charts, and considering $f \circ g^{-1}$ instead of f, together with the invariance of critical points under changes of charts, we may view f as defined on an open subset of $\mathbf{R}^{n-1}$. We can then apply induction again to conclude the proof of our first assertion.

Proof of 2. Again we work locally, and we may view f as defined on the closed n-cube of radius r centered at some point a. We denote this cube by $C_r(a)$. For $m \geqq n/p$, it will suffice to prove that

$$f(Z_m \cap C_r(a))$$

has measure 0. For large N, we cut up each side of the cube into N equal segments, thus obtaining a decomposition of the cube into N^n small cubes. By Taylor's formula, if a small cube contains a critical point $x \in Z_m$, then for any point y of this small cube we have

$$|f(y) - f(x)| \leqq K|x - y|^{m+1} \leqq K(2r/N)^{m+1},$$

where K is a bound for the derivatives of f up to order $m + 1$, and we use the sup norm. Hence the image of Z_m contained in a small cube is itself contained in a cube whose radius is given by the right-hand side, and whose volume in R^p is therefore bounded by

$$K^p(2r/N)^{p(m+1)}.$$

We have n at most N^n such images to consider, and we therefore see that

$$f(Z_m \cap C_r(a))$$

is contained in a union of cubes in R^p, the sum of whose volumes is bounded by

$$K^p N^n (2r/N)^{p(m+1)} \leqq K^p (2r)^{p(m+1)} N^{n-p(m+1)}.$$

Since $m \geqq n/p$, we see that the right-hand side of this estimate behaves like $1/N$ as N becomes large, and hence that the union of the cubes in R^p has arbitrarily small measure, thereby proving Sard's theorem.

Sard's theorem is harder to prove in the case f is C^p with finite p [29], but $p = \infty$ already is quite useful.

§2. Change of variables formula

We first deal with the simplest of cases. We consider vectors $v_1, \ldots, v_n$ in R^n and we define the **block** B spanned by these vectors to be the set of points

$$t_1 v_1 + \cdots + t_n v_n$$

with $0 \leqq t_i \leqq 1$. We say that the block is **degenerate** (in R^n) if the vectors $v_1, \ldots, v_n$ are linearly dependent. Otherwise, we say that the block is **non-degenerate**, or is a **proper block** in R^n.

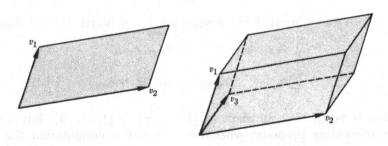

We see that a block in $\mathbf{R}^2$ is nothing but a parallelogram, and a block in $\mathbf{R}^3$ is nothing but a parallelepiped (when not degenerate).

We shall sometimes use the word volume instead of measure when applied to blocks or their images under maps, for the sake of geometry.

We denote by $\mathrm{Vol}\,(v_1, \ldots, v_n)$ the volume of the block B spanned by $v_1, \ldots, v_n$. We define the **oriented volume**

$$\mathrm{Vol}^0\,(v_1, \ldots, v_n) \,=\, \pm\,\mathrm{Vol}\,(v_1, \ldots, v_n),$$

taking the $+$ sign if $\mathrm{Det}\,(v_1, \ldots, v_n) > 0$ and the $-$ sign if

$$\mathrm{Det}\,(v_1, \ldots, v_n) < 0.$$

The determinant is viewed as the determinant of the matrix whose column vectors are $v_1, \ldots, v_n$, in that order.

We recall the following characterization of determinants. Suppose that we have a product

$$(v_1, \ldots, v_n) \mapsto v_1 \wedge v_2 \wedge \cdots \wedge v_n$$

which to each n-tuple of **vectors** associates a number, such that the product is multilinear, alternating, and such that

$$e_1 \wedge \cdots \wedge e_n = 1$$

if $e_1, \ldots, e_n$ are the unit vectors. Then this product is necessarily the determinant, that is, it is uniquely determined. "Alternating" means that if $v_i = v_j$ for some $i \neq j$, then

$$v_1 \wedge \cdots \wedge v_n = 0.$$

The uniqueness is easily proved, and we recall this short proof. We can write

$$v_i = a_{i1}e_1 + \cdots + a_{in}e_n$$

for suitable numbers a_{ij}, and then

$$v_1 \wedge \cdots \wedge v_n = (a_{11}e_1 + \cdots + a_{1n}e_n) \wedge \cdots \wedge (a_{n1}e_1 + \cdots + a_{nn}e_n)$$

$$= \sum_\sigma a_{1,\sigma(1)}e_{\sigma(1)} \wedge \cdots \wedge a_{n,\sigma(n)}e_{\sigma(n)}$$

$$= \sum_\sigma a_{1,\sigma(1)} \cdots a_{n,\sigma(n)}e_{\sigma(1)} \wedge \cdots \wedge e_{\sigma(n)}.$$

The sum is taken over all maps $\sigma: \{1, \ldots, n\} \to \{1, \ldots, n\}$, but because of the alternating property, whenever σ is not a permutation the term corresponding to σ is equal to 0. Hence the sum may be taken only over all permutations. Since

$$e_{\sigma(1)} \wedge \cdots \wedge e_{\sigma(n)} = \varepsilon(\sigma)e_1 \wedge \cdots \wedge e_n$$

where $\varepsilon(\sigma) = 1$ or -1 is a sign depending only on σ, it follows that the alternating product is completely determined by its value $e_1 \wedge \cdots \wedge e_n$, and in particular is the determinant if this value is equal to 1.

Proposition 1. *We have*

$$\mathrm{Vol}^0 (v_1, \ldots, v_n) = \mathrm{Det}\, (v_1, \ldots, v_n)$$

and

$$\mathrm{Vol}\, (v_1, \ldots, v_n) = |\mathrm{Det}\, (v_1, \ldots, v_n)|.$$

Proof. If $v_1, \ldots, v_n$ are linearly dependent, then the determinant is equal to 0, and the volume is also equal to 0, for instance by Lemma 3 of §1. So our formula holds in this case. It is clear that

$$\mathrm{Vol}^0 (e_1, \ldots, e_n) = 1.$$

To show that Vol^0 satisfies the characteristic properties of the determinant, all we have to do now is to show that it is linear in each variable, say the first. In other words, we must prove

(*) $\mathrm{Vol}^0 (cv, v_2, \ldots, v_n) = c\, \mathrm{Vol}^0 (v, v_2, \ldots, v_n)$ for $c \in \mathbf{R}$,

(**) $\mathrm{Vol}^0 (v + w, v_2, \ldots, v_n)$

$$= \mathrm{Vol}^0 (v, v_2, \ldots, v_n) + \mathrm{Vol}^0 (w, v_2, \ldots, v_n).$$

As to the first assertion, suppose first that c is some positive integer k. Let B be the block spanned by $v, v_2, \ldots, v_n$. We may assume without loss of generality that $v, v_2, \ldots, v_n$ are linearly independent (otherwise, the relation is obviously true, both sides being equal to 0). We verify at once from the definition that if $B(v, v_2, \ldots, v_n)$ denotes the block spanned by $v, v_2, \ldots, v_n$ then $B(kv, v_2, \ldots, v_n)$ is the union of the two sets

$$B((k-1)v, v_2, \ldots, v_n) \quad \text{and} \quad B(v, v_2, \ldots, v_n) + (k-1)v$$

which have only a set of measure 0 in common, as one verifies at once from the definitions.

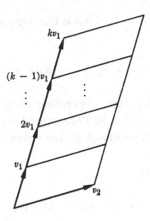

Therefore, we find that

$$\text{Vol}\,(kv, v_2, \ldots, v_n) = \text{Vol}\,((k-1)v, v_2, \ldots, v_n) + \text{Vol}\,(v, v_2, \ldots, v_n)$$
$$= (k-1)\,\text{Vol}\,(v, v_2, \ldots, v_n) + \text{Vol}\,(v, v_2, \ldots, v_n)$$
$$= k\,\text{Vol}\,(v, v_2, \ldots, v_n),$$

as was to be shown.

Now let

$$v = v_1/k$$

for a positive integer k. Then applying what we have just proved shows that

$$\text{Vol}\left(\frac{1}{k}\,v_1, v_2, \ldots, v_n\right) = \frac{1}{k}\,\text{Vol}\,(v_1, \ldots, v_n).$$

Writing a positive rational number in the form $m/k = m \cdot 1/k$, we conclude that the first relation holds when c is a positive rational number. If r is a positive real number, we find positive rational numbers c, c' such that $c \leq r \leq c'$. Since

$$B(cv, v_2, \ldots, v_n) \subset B(rv, v_2, \ldots, v_n) \subset B(c'v, v_2, \ldots, v_n),$$

we conclude that

$$c\,\text{Vol}\,(v, v_2, \ldots, v_n) \leq \text{Vol}\,(rv, v_2, \ldots, v_n) \leq c'\,\text{Vol}\,(v, v_2, \ldots, v_n).$$

Letting c, c' approach r as a limit, we conclude that for any real number $r \geq 0$ we have

$$\text{Vol}\,(rv, v_2, \ldots, v_n) = r\,\text{Vol}\,(v, v_2, \ldots, v_n).$$

Finally, we note that $B(-v, v_2, \ldots, v_n)$ is the translation of

$$B(v, v_2, \ldots, v_n)$$

by $-v$ so that these two blocks have the same volume. This proves the first assertion.

As for the second, we look at the geometry of the situation, which is made clear by the following picture in case $v = v_1, w = v_2$.

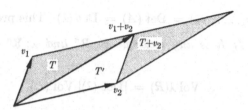

The block spanned by $v_1, v_2, \ldots$ consists of two "triangles" T, T' having only a set of measure zero in common. The block spanned by $v_1 + v_2$ and v_2 consists of T' and the translation $T + v_2$. It follows that these two blocks have the same volume. We conclude that for any number c,

$$\mathrm{Vol}^0\,(v_1 + cv_2, v_2, \ldots, v_n) = \mathrm{Vol}^0\,(v_1, v_2, \ldots, v_n).$$

Indeed, if $c = 0$ this is obvious, and if $c \neq 0$ then

$$c\,\mathrm{Vol}^0\,(v_1 + cv_2, v_2) = \mathrm{Vol}^0\,(v_1 + cv_2, cv_2)$$

$$= \mathrm{Vol}^0\,(v_1, cv_2) = c\,\mathrm{Vol}^0\,(v_1, v_2).$$

We can then cancel c to get our conclusion.

To prove the linearity of Vol^0 with respect to its first variable, we may assume that $v_2, \ldots, v_n$ are linearly independent, otherwise both sides of (**) are equal to 0. Let v_1 be so chosen that $\{v_1, \ldots, v_n\}$ is a basis of $\mathbf{R}^n$. Then by induction, and what has been proved above,

$$\mathrm{Vol}^0\,(c_1v_1 + \cdots + c_nv_n, v_2, \ldots, v_n)$$

$$= \mathrm{Vol}^0\,(c_1v_1 + \cdots + c_{n-1}v_{n-1}, v_2, \ldots, v_n)$$

$$= \mathrm{Vol}^0\,(c_1v_1, v_2, \ldots, v_n)$$

$$= c_1\,\mathrm{Vol}^0\,(v_1, \ldots, v_n).$$

From this the linearity follows at once, and the theorem is proved.

Corollary 1. *Let S be the unit cube spanned by the unit vectors in $\mathbf{R}^n$. Let $\lambda\colon \mathbf{R}^n \to \mathbf{R}^n$ be a linear map. Then*

$$\text{Vol } \lambda(S) = |\text{Det }(\lambda)|.$$

Proof. If $v_1, \ldots, v_n$ are the images of $e_1, \ldots, e_n$ under λ, then $\lambda(S)$ is the block spanned by $v_1, \ldots, v_n$. If we represent λ by the matrix $A = (a_{ij})$, then

$$v_i = a_{1i}e_1 + \cdots + a_{ni}e_n,$$

and hence $\text{Det }(v_1, \ldots, v_n) = \text{Det }(A) = \text{Det }(\lambda)$. This proves the corollary.

Corollary 2. *If R is any rectangle in $\mathbf{R}^n$ and $\lambda\colon \mathbf{R}^n \to \mathbf{R}^n$ is a linear map, then*

$$\text{Vol } \lambda(R) = |\text{Det }(\lambda)| \text{ Vol }(R).$$

Proof. After a translation, we can assume that the rectangle is a block. If $R = \lambda_1(S)$ where S is the unit cube, then

$$\lambda(R) = \lambda \circ \lambda_1(S),$$

whence by Corollary 1

$$\text{Vol } \lambda(R) = |\text{Det }(\lambda \circ \lambda_1)| = |\text{Det }(\lambda) \text{ Det }(\lambda_1)| = |\text{Det }(\lambda)| \text{ Vol }(R).$$

The next theorem extends Corollary 2 to the more general case where the linear map λ is replaced by an arbitrary C^1-invertible map. The proof then consists of replacing the linear map by its derivative and estimating the error thus introduced. For this purpose, we have the **Jacobian determinant**

$$\Delta_f(x) = \text{Det } J_f(x) = \text{Det } f'(x)$$

where $J_f(x)$ is the Jacobian matrix, and $f'(x)$ is the derivative of the map $f\colon U \to \mathbf{R}^n$.

Proposition 2. *Let R be a rectangle in $\mathbf{R}^n$, contained in some open set U. Let $f\colon U \to \mathbf{R}^n$ be a C^1 map, which is C^1-invertible on U. Then*

$$\mu\big(f(R)\big) = \int_R |\Delta_f| \, d\mu.$$

Proof. When f is linear, this is nothing but Corollary 2 of the preceding theorem. We shall prove the general case by approximating f by its derivative. Let us first assume that R is a cube for simplicity. Given ε, let P be a partition of R, obtained by dividing each side of R into N equal segments for large N. Then R is partitioned into N^n subcubes which we denote by S_j $(j = 1, \ldots, N^n)$. We let a_j be the center of S_j.

We have

$$\operatorname{Vol} f(R) = \sum_j \operatorname{Vol} f(S_j)$$

because the images $f(S_j)$ have only sets of measure 0 in common. We investigate $f(S_j)$ for each j. The derivative f' is uniformly continuous on R. Given ε, we assume that N has been taken so large that for $x \in S_j$ we have

$$f(x) = f(a_j) + \lambda_j(x - a_j) + \varphi(x - a_j),$$

where $\lambda_j = f'(a_j)$ and

$$|\varphi(x - a_j)| \leqq |x - a_j|\varepsilon.$$

To determine $\operatorname{Vol} f(S_j)$ we must therefore investigate $f(S)$ where S is a cube centered at the origin, and f has the form

$$f(x) = \lambda x + \varphi(x), \qquad\qquad |\varphi(x)| \leqq |x|\varepsilon.$$

on the cube S. (We have made suitable translations which don't affect volumes.) We have

$$\lambda^{-1} \circ f(x) = x + \lambda^{-1} \circ \varphi(x),$$

so that $\lambda^{-1} \circ f$ is nearly the identity map. For some constant C, we have for $x \in S$

$$|\lambda^{-1} \circ \varphi(x)| \leqq C\varepsilon.$$

From the lemma after the proof of the inverse mapping theorem, we conclude that $\lambda^{-1} \circ f(S)$ contains a cube of radius

$$(1 - C\varepsilon) \text{ radius } S),$$

and trivial estimates show that $\lambda^{-1} \circ f(S)$ is contained in a cube of radius

$$(1 + C\varepsilon) \text{ (radius } S).$$

We apply λ to these cubes, and determine their volumes. Putting indices j on everything, we find that

$$|\operatorname{Det} f'(a_j)| \operatorname{Vol}(S_j) - \varepsilon C_1 \operatorname{Vol}(S_j)$$

$$\leqq \operatorname{Vol} f(S_j) \leqq |\operatorname{Det} f'(a_j)| \operatorname{Vol}(S_j) + \varepsilon C_1 \operatorname{Vol}(S_j)$$

with some fixed constant C_1. Summing over j and estimating $|\Delta_f|$, we see that our theorem follows at once.

Remark. We assumed for simplicity that R was a cube. Actually, by changing the norm on each side, multiplying by a suitable constant, and taking the sup of the adjusting norms, we see that this involves no loss of generality. Alternatively, we can approximate a given rectangle by cubes.

Corollary. *If g is continuous on $f(R)$, then*

$$\int_{f(R)} g \, d\mu = \int_R (g \circ f)|\Delta_f| \, d\mu.$$

Proof. The functions g and $(g \circ f)|\Delta_f|$ are uniformly continuous on $f(R)$ and R respectively. Let us take a partition of R and let $\{S_j\}$ be the subrectangles of this partition. If δ is the maximum length of the sides of the subrectangles of the partition, then $f(S_j)$ is contained in a rectangle whose sides have length $\leq C \delta$ for some constant C. We have

$$\int_{f(R)} g \, d\mu = \sum_j \int_{f(S_j)} g \, d\mu.$$

The sup and inf of g on $f(S_j)$ differ only by ε if δ is taken sufficiently small. Using the theorem, applied to each S_j, and replacing g by its minimum m_j and maximum M_j on S_j, we see that the corollary follows at once.

Change of variables formula. *Let U be open in $\mathbf{R}^n$ and let $f: U \to \mathbf{R}^n$ be a C^1 map, which is C^1 invertible on U. Let $g \in \mathscr{L}^1(f(U))$. Then $(g \circ f)|\Delta_f|$ is in $\mathscr{L}^1(U)$ and we have*

$$\int_{f(U)} g \, d\mu = \int_U (g \circ f)|\Delta_f| \, d\mu.$$

Proof. Let R be a closed rectangle contained in U. We shall first prove that the restriction of $(g \circ f)|\Delta_f|$ to R is in $\mathscr{L}^1(R)$, and that the formula holds when U is replaced by R. We know that $C_c(f(U))$ is L^1-dense in $\mathscr{L}^1(f(U))$, by *Real Analysis*, Theorem 6 of Chapter XII, §3. Hence there exists a sequence $\{g_k\}$ in $C_c(f(U))$ which is L^1-convergent to g. Using *Real Analysis*, Theorem 4 of Chapter X, §5, we may assume that $\{g_k\}$ converges pointwise to g except on a set Z of measure 0 in $f(U)$. By Lemma 2 of §1, we know that $f^{-1}(Z)$ has measure 0.

Let $g_k^* = (g_k \circ f)|\Delta_f|$. Each function g_k^* is continuous on R. The sequence $\{g_k^*\}$ converges almost everywhere to $(g \circ f)|\Delta_f|$ restricted to R. It is in fact an L^1-Cauchy sequence in $\mathscr{L}^1(R)$. To see this, we have by the result for rectangles and continuous functions (corollary of the preceding theorem):

$$\int_R |g_k^* - g_m^*| \, d\mu = \int_{f(R)} |g_k - g_m| \, d\mu,$$

so the Cauchy nature of the sequence $\{g_k^*\}$ is clear from that of $\{g_k\}$. It follows that the restriction of $(g \circ f)|\Delta_f|$ to R is the L^1-limit of $\{g_k^*\}$, and is in $\mathscr{L}^1(R)$. It also follows that the formula of the theorem holds for R, that is

$$\int_{f(A)} g \, d\mu = \int_A (g \circ f)|\Delta_f| \, d\mu$$

when $A = R$.

The theorem is now seen to hold for any measurable subset A of R, since $f(A)$ is measurable, and since a function g in $\mathscr{L}^1(f(A))$ can be extended to a function in $\mathscr{L}^1(f(R))$ by giving it the value 0 outside $f(A)$. From this it follows that the theorem holds if A is a finite union of rectangles contained in U. We can find a sequence of rectangles $\{R_m\}$ contained in U whose union is equal to U, because U is separable. Taking the usual stepwise complementation, we can find a disjoint sequence of measurable sets

$$A_m = R_m - (R_1 \cup \cdots \cup R_{m-1})$$

whose union is U, and such that our theorem holds if $A = A_m$. Let

$$h_m = g_{f(A_m)} = g\chi_{f(A_m)} \quad \text{and} \quad h_m^* = (h_m \circ f)|\Delta_f|.$$

Then $\sum h_m$ converges to g and $\sum h_m^*$ converges to $(g \circ f)|\Delta_f|$. Our theorem follows from Corollary 5 of the dominated convergence theorem in *Real Analysis*.

Note. In dealing with polar coordinates or the like, one sometimes meets a map f which is invertible except on a set of measure 0, e.g. the polar coordinate map. It is now trivial to recover a result covering this type of situation.

Corollary. *Let U be open in $\mathbf{R}^n$ and let $f: U \to \mathbf{R}^n$ be a C^1 map. Let A be a measurable subset of U such that the boundary of A has measure 0, and such that f is C^1 invertible on the interior of A. Let g be in $\mathscr{L}^1(f(A))$. Then $(g \circ f)|\Delta_f|$ is in $\mathscr{L}^1(A)$ and*

$$\int_{f(A)} g \, d\mu = \int_A (g \circ f)|\Delta_f| \, d\mu.$$

Proof. Let U_0 be the interior of A. The sets $f(A)$ and $f(U_0)$ differ only by a set of measure 0, namely $f(\partial A)$. Also the sets A, U_0 differ only by a set of measure 0. Consequently we can replace the domains of integration $f(A)$ and A by $f(U_0)$ and U_0, respectively. The theorem applies to conclude the proof of the corollary.

§3. *Orientation*

Let U, V be open sets in half spaces of $\mathbf{R}^n$ and let $\varphi: U \to V$ be a C^1 isomorphism. We shall say that φ is **orientation preserving** if the Jacobian determinant $\Delta_\varphi(x)$ is > 0, all $x \in U$. If the Jacobian determinant is negative, then we say that φ is orientation **reversing**.

Let X be a C^p manifold, $p \geq 1$, and let $\{(U_i, \varphi_i)\}$ be an atlas. We say that this atlas is **oriented** if all transition maps $\varphi_j \circ \varphi_i^{-1}$ are orientation preserving. Two atlases $\{(U_i, \varphi_i)\}$ and $\{(V_\alpha, \psi_\alpha)\}$ are said to **define the same orientation**, or to be **orientation equivalent**, if their union is oriented. We can also define locally a chart (V, ψ) to be **orientation compatible** with the oriented atlas $\{(U_i, \varphi_i)\}$ if all transition maps $\varphi_i \circ \varphi^{-1}$ (defined whenever $U_i \cap V$ is not empty) are orientation preserving. An orientation equivalence class of oriented atlases is said to define an **oriented** manifold, or to be an **orientation** of the manifold. It is a simple exercise to verify that if a manifold has an orientation, then it has two distinct orientations.

The standard examples of the Moebius strip or projective plane show that not all manifolds admit orientations. We shall now see that the boundary of an oriented manifold with boundary can be given a natural orientation.

Let $\varphi: U \to \mathbf{R}^n$ be an oriented chart at a boundary point of X, such that

(1) *If $(x_1, \ldots, x_n)$ are the local coordinates of the chart, then the boundary points correspond to those points in $\mathbf{R}^n$ satisfying $x_1 = 0$, and*

(2) *The points of U not in the boundary have coordinates satisfying $x_1 < 0$.*

Then $(x_2, \ldots, x_n)$ are the local coordinates for a chart of the boundary, namely the restriction of φ to $\partial X \cap U$, and the picture is as follows.

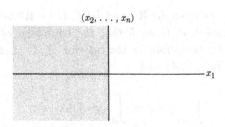

We may say that we have considered a chart φ such that the manifold lies to the left of its boundary. If the reader thinks of a domain in $\mathbf{R}^2$, having a smooth curve for its boundary, as on the following picture, he will see that our choice of chart corresponds to what is usually visualized as "counterclockwise" orientation.

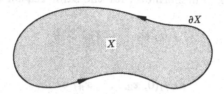

The collection of all pairs $(U \cap \partial X, \varphi \mid (U \cap \partial X))$, *chosen according to the criteria described above, is obviously an atlas for the boundary ∂X, and we contend that it is an oriented atlas.*

We prove this easily as follows. If

$$(x_1, \ldots, x_n) = x \quad \text{and} \quad (y_1, \ldots, y_n) = y$$

are coordinate systems at a boundary point corresponding to choices of charts made according to our specifications, then we can write $y = f(x)$ where $f = (f_1, \ldots, f_n)$ is the transition mapping. Since we deal with oriented charts for X, we know that $\Delta_f(x) > 0$ for all x. Since f maps boundary into boundary, we have

$$f_1(0, x_2, \ldots, x_n) = 0$$

for all $x_2, \ldots, x_n$. Consequently the Jacobian matrix of f at a point $(0, x_2, \ldots, x_n)$ is equal to

$$\begin{pmatrix} D_1 f_1(0, x_2, \ldots, x_n) & 0 \cdots\cdots 0 \\ * & \\ * & \Delta_g^{(n-1)} \\ * & \end{pmatrix}$$

where $\Delta_g^{(n-1)}$ is the Jacobian matrix of the transition map g induced by f on the boundary, and given by

$$y_2 = f_2(0, x_2, \ldots, x_n)$$
$$\vdots \qquad\qquad \vdots$$
$$y_n = f_n(0, x_2, \ldots, x_n).$$

However, we have

$$D_1 f_1(0, x_2, \ldots, x_n) = \lim_{h \to 0} \frac{f_1(h, x_2, \ldots, x_n)}{h},$$

taking the limit with $h < 0$ since by prescription, points of X have coordinates with $x_1 < 0$. Furthermore, for the same reason we have

$$f_1(h, x_2, \ldots, x_n) < 0.$$

Consequently

$$D_1 f_1(0, x_2, \ldots, x_n) > 0.$$

From this it follows that $\Delta_g^{(n-1)}(x_2, \ldots, x_n) > 0$, thus proving our assertion that the atlas we have defined for ∂X is oriented.

From now on, when we deal with an oriented manifold, it is understood that its boundary is taken with orientation described above, and called the induced orientation.

§4. The measure associated with a differential form

Let X be a manifold of class C^p with $p \geqq 1$. We assume from now on that X is Hausdorff and has a countable base. Then we know that X admits C^p partitions of unity, subordinated to any given open covering.

(Actually, instead of the conditions we assumed, we could just as well have assumed the existence of C^p partitions of unity, which is the precise condition to be used in the sequel.)

We can define the **support** of a differential form as we define the support of a function. It is the closure of the set of all $x \in X$ such that $\omega(x) \neq 0$. If ω is a form of class C^q and α is a C^q function on X, then we can form the product $\alpha\omega$, which is the form whose value at x is $\alpha(x)\omega(x)$. If α has compact support, then $\alpha\omega$ has compact support. Later, we shall study the integration of forms, and reduce this to a local problem by means of partitions of unity, in which we multiply a form by functions.

We assume that the reader is familiar with the correspondence between certain functionals on continuous functions with compact support and measures. Cf. *Real Analysis* for this. We just recall some terminology.

We denote by $C_c(X)$ the vector space of continuous functions on X with **compact support** (i.e. vanishing outside a compact set). We write $C_c(X, \mathbf{R})$ or $C_c(X, \mathbf{C})$ if we wish to distinguish between the real or complex valued functions.

We denote by $C_K(X)$ the subspace of $C_c(X)$ consisting of those functions which vanish outside K. (Same notation $C_S(X)$ for those functions which are 0 outside any subset S of X. Most of the time, the useful subsets in this context are the compact subsets K.)

A linear map λ of $C_c(X)$ into the complex numbers (or into a normed vector space, for that matter) is said to be **bounded** if there exists some $C \geq 0$ such that we have

$$|\lambda f| \leq C\|f\|$$

for all $f \in C_c(X)$. Thus λ is bounded if and only if λ is continuous for the norm topology.

A linear map λ of $C_c(X)$ into the complex numbers is said to be **positive** if we have $\lambda f \geq 0$ whenever f is real and ≥ 0.

Lemma 1. *Let* $\lambda\colon C_c(X) \to \mathbf{C}$ *be a positive linear map. Then* λ *is bounded on* $C_K(X)$ *for any compact* K.

Proof. By the corollary of Urysohn's lemma, there exists a continuous real function $g \geq 0$ on X which is 1 on K and has compact support. If $f \in C_K(X)$, let $b = \|f\|$. Say f is real. Then $bg \pm f \geq 0$, whence

$$\lambda(bg) \pm \lambda f \geq 0$$

and $|\lambda f| \leq b\lambda(g)$. Thus λg is our desired bound.

A complex valued linear map on $C_c(X)$ which is bounded on each subspace $C_K(X)$ for every compact K will be called a **C_c-functional** on $C_c(X)$, or more simply, a **functional**. A functional on $C_c(X)$ which is also continuous for the sup norm will be called a **bounded** functional. It is clear that a bounded functional is also a C_c-functional.

Lemma 2. *Let* $\{W_\alpha\}$ *be an open covering of* X. *For each index* α, *let* λ_α *be a functional on* $C_c(W_\alpha)$. *Assume that for each pair of indices* α, β *the functionals* λ_α *and* λ_β *are equal on* $C_c(W_\alpha \cap W_\beta)$. *Then there exists a unique functional* λ *on* X *whose restriction to each* $C_c(W_\alpha)$ *is equal to* λ_α. *If each* λ_α *is positive, then so is* λ.

Proof. Let $f \in C_c(X)$ and let K be the support of f. Let $\{h_i\}$ be a partition of unity over K subordinated to a covering of K by a finite number of the open sets W_α. Then each $h_i f$ has support in some $W_{\alpha(i)}$ and we define

$$\lambda f = \sum_i \lambda_{\alpha(i)}(h_i f).$$

We contend that this sum is independent of the choice of $\alpha(i)$, and also of the choice of partition of unity. Once this is proved, it is then obvious that λ is a functional which satisfies our requirements. We now prove this independence. First note that if $W_{\alpha'(i)}$ is another one of the open sets W_α in which the support of $h_i f$ is contained, then $h_i f$ has support in the intersection $W_{\alpha(i)} \cap W_{\alpha'(i)}$, and our assumption concerning our functionals λ_α shows that the corresponding term in the sum does not depend on the choice of index $\alpha(i)$. Next, let $\{g_k\}$ be another partition of unity over K subordinated to some covering of K by a finite number of the open sets W_α. Then for each i,

$$h_i f = \sum_k g_k h_i f,$$

whence

$$\sum_i \lambda_{\alpha(i)}(h_i f) = \sum_i \sum_k \lambda_{\alpha(i)}(g_k h_i f).$$

If the support of $g_k h_i f$ is in some W_α, then the value $\lambda_\alpha(g_k h_i f)$ is independent of the choice of index α. The expression on the right is then symmetric with respect to our two partitions of unity, whence our theorem follows.

Theorem 1. *Let* $\dim X = n$ *and let* ω *be an n-form on X of class C^0, that is continuous. Then there exists a unique positive functional λ on $C_c(X)$ having the following property. If (U, φ) is a chart and*

$$\omega(x) = f(x)\, dx_1 \wedge \cdots \wedge dx_n$$

is the local representation of ω in this chart, then for any $g \in C_c(X)$ with support in U, we have

(1) $$\lambda g = \int_{\varphi U} g_\varphi(x) |f(x)|\, dx,$$

where g_φ represents g in the chart [i.e. $g_\varphi(x) = g(\varphi^{-1}(x))$], and dx is Lebesgue measure.

Proof. The integral in (1) defines a positive functional on $C_c(U)$. The change of variables formula shows that if (U, φ) and (V, ψ) are two charts, and if g has support in $U \cap V$, then the value of the functional is independent of the choice of charts. Thus we get a positive functional by the general localization lemma for functionals.

The positive measure corresponding to the functional in Theorem 1 will be called the **measure associated with** $|\omega|$, and can be denoted by $\mu_{|\omega|}$.

Theorem 1 does not need any orientability assumption. With such an assumption, we have a similar theorem, obtained without taking the absolute value.

Theorem 2. *Let* $\dim X = n$ *and assume that* X *is oriented. Let* ω *be an* n-*form on* X *of class* C^0. *Then there exists a unique functional* λ *on* $C_c(X)$ *having the following property. If* (U, φ) *is an oriented chart and*

$$\omega(x) = f(x)\, dx_1 \wedge \cdots \wedge dx_n$$

is the local representation of ω *in this chart, then for any* $g \in C_c(X)$ *with support in* U, *we have*

$$\lambda g = \int_{\varphi U} g_\varphi(x) f(x)\, dx,$$

where g_φ *represents* g *in the chart, and* dx *is Lebesgue measure.*

Proof. Since the Jacobian determinant of transition maps belonging to oriented charts is positive, we see that Theorem 2 follows like Theorem 1 from the change of variables formula (in which the absolute value sign now becomes unnecessary) and the existence of partitions of unity.

If λ is the functional of Theorem 2, we shall call it the functional associated with ω. For any function $g \in C_c(X)$, we define

$$\int_X g\omega = \lambda g.$$

If in particular ω has compact support, we can also proceed directly as follows. Let $\{\alpha_i\}$ be a partition of unity over X such that each α_i has compact support. We define

$$\int_X \omega = \sum_i \int_X \alpha_i \omega,$$

all but a finite number of terms in this sum being equal to 0. As usual, it is immediately verified that this sum is in fact independent of the choice of partition of unity, and in fact, we could just as well use only a partition of unity over the support of ω. Alternatively, if α is a function in $C_c(X)$ which is equal to 1 on the support of ω, then we could also define

$$\int_X \omega = \int_X \alpha\omega.$$

It is clear that these two possible definitions are equivalent.

For an interesting theorem at the level of this chapter, see J. Moser's paper "On the volume element on a manifold," *Transactions AMS* **120** (December 1965) pp. 286–294.

CHAPTER IX

Stokes' Theorem

Throughout the chapter, all manifolds are assumed finite dimensional. They may have a boundary.

§1. Stokes' theorem for a rectangular simplex

If X is a manifold and Y a submanifold, then any differential form on X induces a form on Y. We can view this as a very special case of the inverse image of a form, under the embedding (injection) map

$$id: Y \to X.$$

In particular, if Y has dimension $n - 1$, and if $(x_1, \ldots, x_n)$ is a system of coordinates for X at some point of Y such that the points of Y correspond to those coordinates satisfying $x_j = c$ for some fixed number c, and index j, and if the form on X is given in terms of these coordinates by

$$\omega(x) = f(x_1, \ldots, x_n) \, dx_1 \wedge \cdots \wedge dx_n,$$

then the restriction of ω to Y (or the form induced on Y) has the representation

$$f(x_1, \ldots, c, \ldots, x_n) \, dx_1 \wedge \cdots \wedge \widehat{dx_j} \wedge \cdots \wedge dx_n.$$

We should denote this induced form by ω_Y, although occasionally we omit the subscript Y. We shall use such an induced form especially when Y is the boundary of a manifold X.

Let

$$R = [a_1, b_1] \times \cdots \times [a_n, b_n]$$

be a rectangle in n-space, that is a product of n closed intervals. The set theoretic boundary of R consists of the union over all $i = 1, \ldots, n$ of the pieces

$$R_i^0 = [a_1, b_1] \times \cdots \times \{a_i\} \times \cdots \times \{a_n, b_n\}$$

$$R_i^1 = [a_1, b_1] \times \cdots \times \{b_i\} \times \cdots \times [a_n, b_n].$$

If

$$\omega(x_1, \ldots, x_n) = f(x_1, \ldots, x_n) \, dx_1 \wedge \cdots \wedge \widehat{dx_j} \wedge \cdots \wedge dx_n$$

is an $(n - 1)$-form, and the roof over anything means that this thing is to be omitted, then we define

$$\int_{R^0} \omega = \int_{a_i}^{b_1} \cdots \int_{a_i}^{\widehat{b_i}} \cdots \int_{a_n}^{b_n} f(x_1, \ldots, a_i, \ldots, x_n) \, dx_1 \cdots \widehat{dx_j} \cdots dx_n,$$

if $i = j$, and 0 otherwise. And similarly for the integral over R_i^1. We define the integral over the oriented boundary to be

$$\int_{\partial^0 R} = \sum_{i=1}^{n} (-1)^i \left[\int_{R_i^0} - \int_{R_i^1} \right].$$

Stokes' theorem for rectangles. *Let R be a rectangle in an open set U in n-space. Let ω be an $(n - 1)$-form on U. Then*

$$\int_R d\omega = \int_{\partial^0 R} \omega.$$

Proof. In two dimensions, the picture looks like this:

It suffices to prove the assertion when ω is a decomposable form, say

$$\omega(x) = f(x_1, \ldots, x_n) \, dx_1 \wedge \cdots \wedge \widehat{dx_j} \wedge \cdots \wedge dx_n.$$

We then evaluate the integral over the boundary of R. If $i \neq j$, then it is clear that

$$\int_{R_i^0} \omega = 0 = \int_{R_i^1} \omega,$$

so that

$$\int_{\partial^0 R} \omega$$

$$= (-1)^j \int_{a_1}^{b_1} \cdots \widehat{\int_{a_j}^{b_j}} \cdots \int_{a_n}^{b_n} [f(x_1, \ldots, a_j, \ldots, x_n)$$

$$- f(x_1, \ldots, b_j, \ldots, x_n)] \, dx_1 \cdots \widehat{dx_j} \cdots dx_n.$$

On the other hand, from the definitions we find that

$$d\omega(x) = \left(\frac{\partial f}{\partial x_1} dx_1 + \cdots + \frac{\partial f}{\partial x_n} dx_n \right) \wedge dx_1 \wedge \cdots \wedge \widehat{dx_j} \wedge \cdots \wedge dx_n$$

$$= (-1)^{j-1} \frac{\partial f}{\partial x_j} dx_1 \wedge \cdots \wedge dx_n.$$

(The $(-1)^{j-1}$ comes from interchanging dx_j with $dx_1, \ldots, dx_{j-1}$. All other terms disappear by the alternation rule.)

Integrating $d\omega$ over R, we may use repeated integration and integrate $\partial f / \partial x_j$ with respect to x_j first. Then the fundamental theorem of calculus for one variable yields

$$\int_{a_j}^{b_j} \frac{\partial f}{\partial x_j} dx_j = f(x_1, \ldots, b_j, \ldots, x_n) - f(x_1, \ldots, a_j, \ldots, x_n).$$

We then integrate with respect to the other variables, and multiply by $(-1)^{j-1}$. This yields precisely the value found for the integral of ω over the oriented boundary $\partial^0 R$, and proves the theorem.

Remark. Stokes' theorem for a rectangle extends at once to a version in which we parametrize a subset of some space by a rectangle. Indeed, if $\sigma : R \to V$ is a C^1 map of a rectangle of dimension n into an open set V in $\mathbf{R}^N$, and if ω is an $(n-1)$-form in V, we may define

$$\int_\sigma d\omega = \int_R \sigma^* d\omega.$$

One can define

$$\int_{\partial\sigma} \omega = \int_{\partial^0 R} \sigma^* \omega,$$

and then we have a formula

$$\boxed{\int_\sigma d\omega = \int_{\partial\sigma} \omega.}$$

In the next section, we prove a somewhat less formal result.

§2. Stokes' theorem on a manifold

Theorem 1. *Let X be an oriented manifold of class C^2, dimension n, and let ω be an $(n-1)$-form on X, of class C^1. Assume that ω has compact support. Then*

$$\int_X d\omega = \int_{\partial X} \omega.$$

Proof. Let $\{\alpha_i\}_{i \in I}$ be a partition of unity, of class C^2. Then

$$\sum_{i \in I} \alpha_i \omega = \omega,$$

and this sum has only a finite number of non-zero terms since the support of ω is compact. Using the additivity of the operation d, and that of the integral, we find

$$\int_X d\omega = \sum_{i \in I} \int_X d(\alpha_i \omega).$$

Suppose that α_i has compact support in some open set V_i of X and that we can prove

$$\int_{V_i} d(\alpha_i \omega) = \int_{V_i \cap \partial X} \alpha_i \omega,$$

in other words we can prove Stokes' theorem locally in V_i. We can write

$$\int_{V_i \cap \partial X} \alpha_i \omega = \int_{\partial X} \alpha_i \omega,$$

and similarly

$$\int_{V_i} d(\alpha_i \omega) = \int_X d(\alpha_i \omega).$$

Using the additivity of the integral once more, we get

$$\int_X d\omega = \sum_{i \in I} \int_X d(\alpha_i \omega) = \sum_{i \in I} \int_{\partial X} \alpha_i \omega = \int_{\partial X} \omega,$$

which yields Stokes' theorem on the whole manifold. Thus our argument with partitions of unity reduces Stokes' theorem to the local case, namely it suffices to prove that for each point of X there exists an open neighborhood V such that if ω has compact support in V, then Stokes' theorem holds with X replaced by V. We now do this.

If the point is not a boundary point, we take an oriented chart (U, φ) at the point, containing an open neighborhood V of the point, satisfying the following conditions: φU is an open ball, and φV is the interior of a rectangle, whose closure is contained in φU. If ω has compact support in V, then its local representation in φU has compact support in φV. Applying Stokes' theorem for rectangles as proved in the preceding section, we find that the two integrals occurring in Stokes' formula are equal to 0 in this case (the integral over an empty boundary being equal to 0 by convention).

Now suppose that we deal with a boundary point. We take an oriented chart (U, φ) at the point, having the following properties. First, φU is described by the following inequalities in terms of local coordinates $(x_1, \ldots, x_n)$:

$$-2 < x_1 \leqq 1 \quad \text{and} \quad -2 < x_j < 2 \quad \text{for} \quad j = 2, \ldots, n.$$

Next, the given point has coordinates $(1, 0, \ldots, 0)$, and that part of U on the boundary of X, namely $U \cap \partial X$, is given in terms of these coordinates by the equation $x_1 = 1$. We then let V consist of those points whose local coordinates satisfy

$$0 < x_1 \leqq 1 \quad \text{and} \quad -1 < x_j < 1 \quad \text{for} \quad j = 2, \ldots, n.$$

If ω has compact support in V, then ω is equal to 0 on the boundary of the rectangle R equal to the closure of φV, except on the face given by $x_1 = 1$, which defines that part of the rectangle corresponding to $\partial X \cap V$. Thus the support of ω looks like the shaded portion of the following picture.

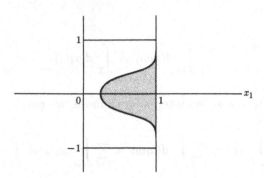

In the sum giving the integral over the boundary of a rectangle as in the previous section, only one term will give a non-zero contribution, corresponding to $i = 1$, which is

$$(-1) \left[\int_{R_1^0} \omega - \int_{R_1^1} \omega \right].$$

Furthermore, the integral over R_1^0 will also be 0, and in the contribution of the integral over R_1^1, the two minus signs will cancel, and yield the integral of ω over the part of the boundary lying in V, because our charts are so chosen that $(x_2, \ldots, x_n)$ is an oriented system of coordinates for the boundary. Thus we find

$$\int_V d\omega = \int_{V \cap \partial X} \omega,$$

which proves Stokes' theorem locally in this case, and concludes the proof of Theorem 1.

For any number of reasons, some of which we consider in the next section, it is useful to formulate conditions under which Stokes' theorem holds even when the form ω does not have compact support. We shall say that ω has **almost compact support** if there exists a decreasing sequence of open sets $\{U_k\}$ in X such that the intersection

$$\bigcap_{k=1}^{\infty} U_k$$

is empty, and a sequence of C^1 functions $\{g_k\}$, having the following properties:

AC 1. *We have $0 \leq g_k \leq 1$, $g_k = 1$ outside U_k, and $g_k\omega$ has compact support.*

AC 2. *If μ_k is the measure associated with $|dg_k \wedge \omega|$ on X, then*

$$\lim_{k \to \infty} \mu_k(\bar{U}_k) = 0.$$

We then have the following application of Stokes' theorem.

Corollary. *Let X be a C^2 oriented manifold, of dimension n, and let ω be an $(n-1)$-form on X, of class C^1. Assume that ω has almost compact support, and that the measures associated with $|d\omega|$ on X and $|\omega|$ on ∂X are finite. Then*

$$\int_X d\omega = \int_{\partial X} \omega.$$

Proof. By our standard form of Stokes' theorem we have

$$\int_{\partial X} g_k\omega = \int_X d(g_k\omega) = \int_X dg_k \wedge \omega + \int_X g_k\, d\omega.$$

We estimate the left-hand side by

$$\left| \int_{\partial X} \omega - \int_{\partial X} g_k\omega \right| = \left| \int_{\partial X} (1 - g_k)\omega \right| \leq \mu_{|\omega|}(U_k \cap \partial X).$$

Since the intersection of the sets U_k is empty, it follows for a purely measure-theoretic reason that

$$\lim_{k \to \infty} \int_{\partial X} g_k\omega = \int_{\partial X} \omega.$$

Similarly,

$$\lim_{k \to \infty} \int_X g_k\, d\omega = \int_X d\omega.$$

The integral of $dg_k \wedge \omega$ over X approaches 0 as $k \to \infty$ by assumption, and the fact that $dg_k \wedge \omega$ is equal to 0 on the complement of $\bar{U}_k$ since g_k is constant on this complement. This proves our corollary.

The above proof shows that the second condition **AC 2** is a very natural one to reduce the integral of an arbitrary form to that of a form with compact support. In the next section, we relate this condition to a question of singularities when the manifold is embedded in some bigger space.

§3. Stokes' theorem with singularities

If X is a compact manifold, then of course every differential form on X has compact support. However, the version of Stokes' theorem which we have given is useful in contexts when we start with an object which is not a manifold, say as a subset of $\mathbf{R}^n$, but is such that when we remove a portion of it, what remains is a manifold. For instance, consider a cone (say the solid cone) as illustrated in the next picture.

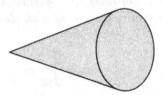

The vertex and the circle surrounding the base disc prevent the cone from being a submanifold of $\mathbf{R}^3$. However, if we delete the vertex and this circle, what remains is a submanifold with boundary embedded in $\mathbf{R}^3$. The boundary consists of the conical shell, and of the base disc (without its surrounding circle). Another example is given by polyhedra, as on the following figure.

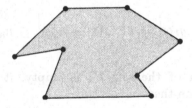

The idea is to approximate a given form by a form with compact support, to which we can apply Theorem 1, and then take the limit. We shall indicate one possible technique to do this.

The word "boundary" has been used in two senses: The sense of point set topology, and the sense of boundary of a manifold. Up to now, they were used in different contexts so no confusion could arise. We must now make a distinction, and therefore use the word boundary only in its manifold sense. If X is a subset of $\mathbf{R}^N$, we denote its closure by $\overline{X}$ as usual. We call the set theoretic difference $\overline{X} - X$ the **frontier** of X in $\mathbf{R}^N$, and denote it by $\mathrm{fr}(X)$.

Let X be a submanifold without boundary of $\mathbf{R}^N$, of dimension n. We know that this means that at each point of X there exists a chart for an open neighborhood of this point in $\mathbf{R}^N$ such that the points of X in this chart correspond to a factor in a product. A point P of $\overline{X} - X$ will be

called a **regular** frontier point of X if there exists a chart at P in $\mathbf{R}^N$ with
local coordinates $(x_1, \ldots, x_N)$ such that P has coordinates $(0, \ldots, 0)$; the
points of X are those with coordinates

$$x_{n+1} = \cdots = x_N = 0 \quad \text{and} \quad x_n < 0;$$

and the points of the frontier of X which lie in the chart are those with
coordinates satisfying

$$x_n = x_{n+1} = \cdots = x_N = 0.$$

The set of all regular frontier points of X will be denoted by ∂X, and will
be called the **boundary** of X. We may say that $X \cup \partial X$ is a submanifold
of $\mathbf{R}^N$, possibly with boundary.

A point of the frontier of X which is not regular will be called **singular**.
It is clear that the set of singular points is closed in $\mathbf{R}^N$. We now formulate
a version of Theorem 1 when ω does not necessarily have compact support
in $X \cup \partial X$. Let S be a subset of $\mathbf{R}^N$. By a **fundamental sequence** of open
neighborhoods of S we shall mean a sequence $\{U_k\}$ of open sets containing
S such that, if W is an open set containing S, then $U_k \subset W$ for all sufficiently
large k.

Let S be the set of singular frontier points of X and let ω be a form
defined on an open neighborhood of $\overline{X}$, and having compact support. The
intersection of supp ω with $(X \cup \partial X)$ need not be compact, so that we
cannot apply Theorem 1 as it stands. The idea is to find a fundamental
sequence of neighborhoods $\{U_k\}$ of S, and a function g_k which is 0 on a
neighborhood of S and 1 outside U_k so that $g_k\omega$ differs from ω only inside
U_k. We can then apply Theorem 1 to $g_k\omega$ and we hope that taking the limit
yields Stokes' theorem for ω itself. However, we have

$$\int_X d(g_k\omega) = \int_X dg_k \wedge \omega + \int_X g_k \, d\omega.$$

Thus we have an extra term on the right, which should go to 0 as $k \to \infty$
if we wish to apply this method. In view of this, we make the following
definition.

Let S be a closed subset of $\mathbf{R}^N$. We shall say that S is **negligible for** X
if there exists an open neighborhood U of S in $\mathbf{R}^N$, a fundamental sequence
of open neighborhoods $\{U_k\}$ of S in U, with $\overline{U}_k \subset U$, and a sequence of
C^1 functions $\{g_k\}$, having the following properties.

NEG 1. *We have $0 \leq g_k \leq 1$. Also, $g_k(x) = 0$ for x in some open neighborhood of S, and $g_k(x) = 1$ for $x \notin U_k$.*

NEG 2. *If ω is an $(n-1)$-form of class C^1 on U, and μ_k is the measure associated with $|dg_k \wedge \omega|$ on $U \cap X$, then μ_k is finite for large k, and*

$$\lim_{k \to \infty} \mu_k(U \cap X) = 0.$$

From our first condition, we see that $g_k \omega$ vanishes on an open neighborhood of S. Since $g_k = 1$ on the complement of $\bar{U}_k$, we have $dg_k = 0$ on this complement, and therefore our second condition implies that the measures induced on X near the singular frontier by $|dg_k \wedge \omega|$ (for $k = 1, 2, \ldots$), are concentrated on shrinking neighborhoods and tend to 0 as $k \to \infty$.

Theorem 2 (Stokes' theorem with singularities). *Let X be an oriented, C^3 submanifold without boundary of $\mathbf{R}^N$. Let $\dim X = n$. Let ω be an $(n-1)$-form of class C^1 on an open neighborhood of $\bar{X}$ in $\mathbf{R}^N$, and with compact support. Assume that:*

(i) *If S is the set of singular points in the frontier of X, then $S \cap \operatorname{supp} \omega$ is negligible for X.*

(ii) *The measures associated with $|d\omega|$ on X, and $|\omega|$ on ∂X, are finite.*

Then

$$\int_X d\omega = \int_{\partial X} \omega.$$

Proof. Let U, $\{U_k\}$, and $\{g_k\}$ satisfy conditions **NEG 1** and **NEG 2**. Then $g_k \omega$ is 0 on an open neighborhood of S, and since ω is assumed to have compact support, one verifies immediately that

$$(\operatorname{supp} g_k \omega) \cap (X \cup \partial X)$$

is compact. Thus Theorem 1 is applicable, and we get

$$\int_{\partial X} g_k \omega = \int_X d(g_k \omega) = \int_X dg_k \wedge \omega + \int_X g_k \, d\omega.$$

We have

$$\left| \int_{\partial X} \omega - \int_{\partial X} g_k \omega \right| \leq \left| \int_{\partial X} (1 - g_k) \omega \right|$$

$$\leq \int_{U_k \cap \partial X} 1 \, d\mu_{|\omega|} = \mu_{|\omega|}(U_k \cap \partial X).$$

Since the intersection of all sets $U_k \cap \partial X$ is empty, it follows from purely measure theoretic reasons that the limit of the right-hand side is 0 as $k \to \infty$. Thus

$$\lim_{k \to \infty} \int_{\partial X} g_k \omega = \int_{\partial X} \omega.$$

For similar reasons, we have

$$\lim_{k \to \infty} \int_X g_k \, d\omega = \int_X d\omega.$$

Our second assumption **NEG 2** guarantees that the integral of $dg_k \wedge \omega$ over X approaches 0. This proves our theorem.

We shall now give criteria for a set to be negligible.

Criterion 1. *Let S, T be compact negligible sets for a submanifold X of $\mathbf{R}^N$ (assuming X without boundary). Then the union $S \cup T$ is negligible for X.*

Proof. Let $U, \{U_k\}, \{g_k\}$ and $V, \{V_k\}, \{h_k\}$ be triples associated with S and T respectively as in conditions **NEG 1** and **NEG 2** (with V replacing U and h replacing g when T replaces S). Let

$$W = U \cup V, \qquad W_k = U_k \cup V_k, \qquad \text{and} \qquad f_k = g_k h_k.$$

Then the open sets $\{W_k\}$ form a fundamental sequence of open neighborhoods of $S \cup T$ in W, and **NEG 1** is trivially satisfied. As for **NEG 2**, we have

$$d(g_k h_k) \wedge \omega = h_k \, dg_k \wedge \omega + g_k \, dh_k \wedge \omega,$$

so that **NEG 2** is also trivially satisfied, thus proving our criterion.

Criterion 2. *Let X be an open set, and let S be a compact subset in $\mathbf{R}^n$. Assume that there exists a closed rectangle R of dimension $m \leq n - 2$ and a C^1 map $\sigma : R \to \mathbf{R}^n$ such that $S = \sigma(R)$. Then S is negligible for X.*

Before giving the proof, we make a couple of simple remarks. First, we could always take $m = n - 2$, since any parametrization by a rectangle of dimension $< n - 2$ can be extended to a parametrization by a rectangle of dimension $n - 2$ simply by projecting away two coordinates. Second, by our first criterion, we see that a finite union of sets as described above, that is parametrized smoothly by rectangles of codimension ≥ 2, are negligible. Third, our Criterion 2, combined with the first criterion, shows that negligibility in this case is local, that is we can subdivide a rectangle into small pieces.

We now prove Criterion 2. Composing σ with a suitable linear map, we may assume that R is a unit cube. We cut up each side of the cube into k equal segments and thus get k^m small cubes. Since the derivative of σ is bounded on a compact set, the image of each small cube is contained in an n-cube in $\mathbf{R}^N$ of radius $\leq C/k$ (by the mean value theorem), whose n-dimensional volume is $\leq (2C)^n/k^n$. Thus we can cover the image by small cubes such that the sum of their n-dimensional volumes is

$$\leq (2C)^n/k^{n-m} \leq (2C)^n/k^2.$$

Lemma. *Let S be a compact subset of $\mathbf{R}^n$. Let U_k be the open set of points x such that $d(x, S) < 2/k$. There exists a C^∞ function g_k on $\mathbf{R}^N$ which is equal to 0 in some open neighborhood of S, equal to 1 outside U_k, $0 \leq g_k \leq 1$, and such that all partial derivatives of g_k are bounded by $C_1 k$, where C_1 is a constant depending only on n.*

Proof. Let φ be a C^∞ function such that $0 \leq \varphi \leq 1$, and

$$\varphi(x) = 0 \quad \text{if} \quad 0 \leq \|x\| \leq \tfrac{1}{2},$$
$$\varphi(x) = 1 \quad \text{if} \quad 1 \leq \|x\|.$$

We use $\| \ \|$ for the sup norm in $\mathbf{R}^n$. The graph of φ looks like this:

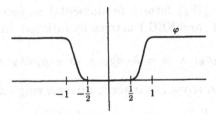

For each positive integer k, let $\varphi_k(x) = \varphi(kx)$. Then each partial derivative $D_i\varphi_k$ satisfies the bound

$$\|D_i\varphi_k\| \leq k\|D_i\varphi\|,$$

which is thus bounded by a constant times k. Let L denote the lattice of integral points in $\mathbf{R}^n$. For each $l \in L$, we consider the function

$$x \mapsto \varphi_k\left(x - \frac{l}{2k}\right).$$

This function has the same shape as φ_k but is translated to the point $l/2k$. Consider the product

$$g_k(x) = \prod \varphi_k\left(x - \frac{l}{2k}\right)$$

taken over all $l \in L$ such that $d(l/2k, S) \leq 1/k$. If x is a point of $\mathbf{R}^n$ such that $d(x, S) < 1/4k$, then we pick an l such that

$$d(x, l/2k) \leq 1/2k.$$

For this l we have $d(l/2, S) < 1/k$, so that this l occurs in the product, and

$$\varphi_k(x - l/2k) = 0.$$

Therefore g_k is equal to 0 in an open neighborhood of S. If, on the other hand, we have $d(x, S) > 2/k$ and if l occurs in the product, that is

$$d(l/2k, S) \leq 1/k,$$

then

$$d(x, l/2k) > 1/k,$$

and hence $g_k(x) = 1$. The partial derivatives of g_k are bounded in the desired manner. This is easily seen, for if x_0 is a point where g_k is not identically 1 in a neighborhood of x_0, then $\|x_0 - l_0/2k\| \leq 1/k$ for some l_0. All other factors $\varphi_k(x - 1/2k)$ will be identically 1 near x_0 unless $\|x_0 - l/2k\| \leq 1/k$. But then $\|l - l_0\| \leq 4$ whence the number of such l is bounded as a function of n (in fact by 9^n). Thus when we take the derivative, we get a sum of at most 9^n terms, each one having a derivative bounded by $C_1 k$ for some constant C_1. This proves our lemma.

We return to the proof of Criterion 2. We observe that when an $(n - 1)$-form ω is expressed in terms of its coordinates,

$$\omega(x) = \sum f_j(x)\, dx_1 \wedge \cdots \wedge \widehat{dx_j} \wedge \cdots \wedge dx_n,$$

then the coefficients f_j are bounded on a compact neighborhood of S. We take U_k as in the lemma. Then for k large, each function

$$x \mapsto f_j(x) D_j g_k(x)$$

is bounded on U_k by a bound $C_2 k$, where C_2 depends on a bound for ω, and on the constant of the lemma. The Lebesgue measure of U_k is bounded by C_3/k^2, as we saw previously. Hence the measure of U_k associated with $|dg_k \wedge \omega|$ is bounded by C_4/k, and tends to 0 as $k \to \infty$. This proves our criterion.

As an example, we now state a simpler version of Stokes' theorem, applying our criteria.

Theorem 3. *Let X be an open subset of $\mathbf{R}^n$. Let S be the set of singular points in the closure of X, and assume that S is the finite union of C^1 images of m-rectangles with $m \leq n - 2$. Let ω be an $(n - 1)$-form defined on an open neighborhood of $\overline{X}$. Assume that ω has compact support, and that the measures associated with $|\omega|$ on ∂X and with $|d\omega|$ on X are finite. Then*

$$\int_X d\omega = \int_{\partial X} \omega.$$

Proof. Immediate from our two criteria and Theorem 2.

We can apply Theorem 3 when, for instance, X is the interior of a poly-hedron, whose interior is open in $\mathbf{R}^n$. When we deal with a submanifold X of dimension n, embedded in a higher dimensional space $\mathbf{R}^N$, then one can reduce the analysis of the singular set to Criterion 2 provided that there exists a finite number of charts for X near this singular set on which the given form ω is bounded. This would for instance be the case with the surface of our cone mentioned at the beginning of the section. Criterion 2 is also the natural one when dealing with manifolds defined by algebraic inequalities. Hironaka tells me that by using the resolution of singularities, one can parametrize a compact set of algebraic singularities as in Criterion 2.

Finally, we note that the condition that ω have compact support in an open neighborhood of $\overline{X}$ is a very mild condition. If for instance X is a bounded open subset of $\mathbf{R}^n$, then $\overline{X}$ is compact. If ω is any form on some open set containing $\overline{X}$, then we can find another form η which is equal to ω on some open neighborhood of $\overline{X}$ and which has compact support. The integrals of η entering into Stokes' formula will be the same as those of ω. To find η, we simply multiply ω with a suitable C^∞ function which is 1 in a neighborhood of $\overline{X}$ and vanishes a little further away. Thus Theorem 3 provides a reasonably useful version of Stokes' theorem which can be applied easily to all the cases likely to arise naturally.

§4. The divergence theorem

Let X be an oriented manifold of dimension n and let Ω be an n-form on X. Let ξ be a vector field on X. Then $d\Omega = 0$, and hence the basic formula for the Lie derivative (Chapter V, §6, Proposition 10) shows that

$$\mathscr{L}_\xi \Omega = d(\Omega \lrcorner \xi).$$

Consequently in this case, Stokes' theorem has the form

$$\int_X \mathscr{L}_\xi \Omega = \int_{\partial X} \Omega \lrcorner \xi.$$

This is called the **divergence theorem**.

Remark. Even if the manifold is not orientable, it is possible to define the notion of density and to formulate a Stokes theorem for densities. Cf. Loomis-Sternberg [13] for the formulation, due to Rasala. However, this formulation reduces at once to a local question (using partitions of unity on densities). Since locally every manifold is orientable, and a density then amounts to a differential form, this more general formulation again reduces to the standard one on an orientable manifold.

Suppose that X is a Riemannian manifold. Then we can define in a natural way an n-form namely at each point x if $v_1, \ldots, v_n$ are in T_x, and are such that

$$\det \langle v_i, v_j \rangle_x$$

is positive, the function

$$\Omega_x (v_1, \ldots, v_n = \sqrt{\det \langle v_i, v_j \rangle_x}$$

is obviously an n-form which is nowhere equal to 0, and will be called the **canonical Riemannian volume form**. If X is orientable, then in each chart compatible with the orientation, this form has a representation

$$\Omega(x) = f(x)\, dx_1 \wedge \cdots \wedge dx_n,$$

and the sign of the corresponding function f in the chart is always positive or always negative.

At a point, the space of n-forms is 1-dimensional. Hence any n-form on a Riemannian manifold can be written as a product $\varphi\Omega$ where φ is a function and Ω is the canonical Riemannian volume form. If ξ is a vector field, and

$$d(\Omega \lrcorner \xi) = \varphi\Omega,$$

with such a function φ, then we call φ the **divergence** of ξ with respect to Ω, or with respect to the Riemannian metric. We denote it by $\operatorname{div}_\Omega \xi$.

By the property of the Lie derivative recalled above, and the present definition, we have in the Riemannian case:

$$\mathscr{L}_\xi \Omega = (\operatorname{div}_\Omega \xi)\Omega.$$

Example. Looking back at the example of Chapter V, §6 we see that if

$$\Omega(x) = dx_1 \wedge \cdots \wedge dx_n$$

is the canonical form on $\mathbf{R}^n$ and ξ is a vector field, then its divergence is given by

$$\mathrm{div}_\Omega\, \xi = \sum_{i=1}^{n} \frac{\partial \xi_i}{\partial x_i}.$$

Let X be an oriented Riemannian manifold. Let ω be the canonical Riemannian volume form on ∂X and let Ω be the canonical Riemannian volume form on X itself. Let $\mathbf{n}_x$ be the unit vector in the tangent space $T_x(X)$ which is perpendicular to $T_x(\partial X)$ and is such that

$$\mathbf{n}_x \wedge \omega(x) = \Omega(x).$$

We shall call $\mathbf{n}_x$ the **unit outward normal vector** to the boundary at x. In an oriented chart, it looks like this.

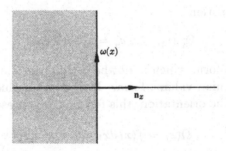

Then by formula **CON 3** of Chapter V, §6 we find

$$\Omega \lrcorner\, \xi = \langle \mathbf{n}, \xi \rangle \omega - \mathbf{n} \wedge (\omega \lrcorner\, \xi),$$

and the restriction of this form to ∂X is simply $\langle \mathbf{n}, \xi \rangle \omega$. Thus we get:

Gauss theorem. *Let X be a Riemannian manifold. Let ω be the canonical Riemannian volume form on ∂X and let Ω be the canonical Riemannian volume form on X itself. Let $\mathbf{n}$ be the unit outward normal vector field to the boundary, and let ξ be a C^1 vector field on X, with compact support. Then*

$$\int_X (\mathrm{div}_\Omega\, \xi)\Omega = \int_{\partial X} \langle \mathbf{n}, \xi \rangle \omega.$$

§5. Cauchy's theorem

It is possible to define a complex analytic (analytic, for short) manifold, using open sets in $\mathbf{C}^n$ and charts such that the transition mappings are analytic. Since analytic maps are C^∞, we see that we get a C^∞ manifold, but with an additional structure, and we call such a manifold (**complex**) **analytic**. It is verified at once that the analytic charts of such a manifold define an orientation.

If $z_1, \ldots, z_n$ are the complex coordinates of $\mathbf{C}^n$, then

$$(z_1, \ldots, z_n, \bar{z}_1, \ldots, \bar{z}_n)$$

can be used as C^∞ local coordinates, viewing $\mathbf{C}^n$ as $\mathbf{R}^{2n}$. If $z_k = x_k + iy_k$, then

$$dz_k = dx_k + idy_k \qquad \text{and} \qquad d\bar{z}_k = dx_k - idy_k.$$

Differential forms can then be expressed in terms of wedge products of the dz_k and $d\bar{z}_k$. For instance

$$dz_k \wedge d\bar{z}_k = 2i\, dy_k \wedge dx_k.$$

The complex standard expression for a differential form is then

$$\omega(z) = \sum_{(i,j)} \varphi_{(i,j)}(z)\, dz_{i_1} \wedge \cdots \wedge dz_{i_r} \wedge d\bar{z}_{j_1} \wedge \cdots \wedge d\bar{z}_{j_s}.$$

Under an analytic change of coordinates, one sees that the numbers r and s remain unchanged, and that if $s = 0$ in one analytic chart, then $s = 0$ in any other analytic chart. Similarly for r. Thus we can speak of a form of type (r, s). A form is said to be **analytic** if $s = 0$, that is if it is of type $(r, 0)$.

We can decompose the exterior derivative d into two components. Namely, we note that if ω is of type (r, s), then $d\omega$ is a sum of forms of type $(r + 1, s)$ and $(r, s + 1)$, say

$$d\omega = (d\omega)_{(r+1,s)} + (d\omega)_{(r,s+1)}.$$

We define

$$\partial\omega = (d\omega)_{(r+1,s)} \qquad \text{and} \qquad \bar{\partial}\omega = (d\omega)_{(r,s+1)}.$$

In terms of local coordinates, it is then easy to verify that if ω is decomposable, and is expressed as

$$\omega(z) = \varphi(z)\, dz_{i_1} \wedge \cdots \wedge dz_{i_r} \wedge d\bar{z}_{j_1} \wedge \cdots \wedge d\bar{z}_{j_s} = \varphi\tilde{\omega},$$

then

$$\partial \omega = \sum \frac{\partial \varphi}{\partial z_k} \, dz_k \wedge \tilde{\omega}$$

and

$$\bar{\partial} \omega = \sum \frac{\partial \varphi}{\partial \bar{z}_k} \, d\bar{z}_k \wedge \tilde{\omega}.$$

In particular, we have

$$\frac{\partial}{\partial z_k} = \frac{1}{2} \left(\frac{\partial}{\partial x_k} - i \frac{\partial}{\partial y_k} \right) \quad \text{and} \quad \frac{\partial}{\partial \bar{z}_k} = \frac{1}{2} \left(\frac{\partial}{\partial x_k} + i \frac{\partial}{\partial y_k} \right).$$

(*Warning*: Note the position of the plus and minus signs in these expressions.)

Thus we have

$$d = \partial + \bar{\partial},$$

and operating with ∂ or $\bar{\partial}$ follows rules similar to the rules for operating with d.

Note that f is analytic if and only if $\bar{\partial} f = 0$. Similarly, we say that a differential form is **analytic** if in its standard expression, the functions $\varphi_{(i,j)}$ are analytic and the form is of type $(r, 0)$, that is there are no $d\bar{z}_j$ present. Equivalently, this amounts to saying that $\bar{\partial} \omega = 0$. The following extension of Cauchy's theorem to several variables is due to Martinelli.

We let $|z|$ be the Euclidean norm,

$$|z| = (z_1 \bar{z}_1 + \cdots + z_n \bar{z}_n)^{1/2}.$$

Cauchy's theorem. *Let f be analytic on an open set in $\mathbf{C}^n$ containing the closed ball of radius R centered at a point ζ. Let*

$$\omega_k(z) = dz_1 \wedge \cdots \wedge dz_n \wedge d\bar{z}_1 \wedge \cdots \wedge \widehat{d\bar{z}_k} \wedge \cdots \wedge d\bar{z}_n$$

and

$$\omega(z) = \sum_{k=1}^{n} (-1)^k \bar{z}_k \omega_k(z).$$

Let S_R be the sphere of radius R centered at ζ. Then

$$f(\zeta) = \varepsilon(n) \frac{(n-1)!}{(2\pi i)^n} \int_{S_R} \frac{f(z)}{|z - \zeta|^{2n}} \, \omega(z - \zeta)$$

where $\varepsilon(n) = (-1)^{n(n+)1/2}$.

Proof. We may assume $\zeta = 0$. First note that

$$\bar{\partial}\omega(z) = \sum_{k=1}^{n} (-1)^k \, d\bar{z}_k \wedge \omega_k(z) = (-1)^{n+1} n \, dz \wedge d\bar{z},$$

where $dz = dz_1 \wedge \cdots \wedge dz_n$ and similarly for $d\bar{z}$. Next, observe that if

$$\psi(z) = \frac{f(z)}{|z|^{2n}} \, \omega(z),$$

then

$$d\psi = 0.$$

This is easily seen. On the one hand, $\partial\psi = 0$ because ω already has $dz_1 \wedge \cdots \wedge dz_n$, and any further dz_1 wedged with this gives 0. On the other hand, since f is analytic, we find that

$$\bar{\partial}\psi(z) = f(z) \, \bar{\partial}\left(\frac{\omega(z)}{|z|^{2n}}\right) = 0$$

by the rule for differentiating a product and a trivial computation.

Therefore, by Stokes' theorem, applied to the annulus between two spheres, for any r with $0 < r \leqq R$ we get

$$\int_{S_R} \psi - \int_{S_r} \psi = 0,$$

or in other words,

$$\int_{S_R} f(z) \, \frac{\omega(z)}{|z|^{2n}} = \int_{S_r} f(z) \, \frac{\omega(z)}{|z|^{2n}}$$

$$= \frac{1}{r^{2n}} \int_{S_r} f(z)\omega(z).$$

Using Stokes' theorem once more, and the fact that $\partial\omega = 0$, we see that this is

$$= \frac{1}{r^{2n}} \int_{B_r} \bar{\partial}(f\omega) = \frac{1}{r^{2n}} \int_{B_r} f \, \bar{\partial}\omega.$$

We can write $f(z) = f(0) + g(z)$, where $g(z)$ tends to 0 as z tends to 0. Thus in taking the limit as $r \to 0$, we may replace f by $f(0)$. Hence our last expression has the same limit as

$$f(0)\frac{1}{r^{2n}}\int_{B_r}\bar{\partial}\omega = f(0)\frac{1}{r^{2n}}\int_{B_r}(-1)^{n+1}n\,dz \wedge d\bar{z}.$$

But

$$dz \wedge d\bar{z} = (-1)^{n(n-1)/2}i^n 2^n\,dy_1 \wedge dx_1 \wedge \cdots \wedge dy_n \wedge dx_n.$$

Interchanging dy_k and dx_k to get the proper orientation gives another contribution of $(-1)^n$, together with the form giving Lebesgue measure. Hence our expression is equal to

$$f(0)(-1)^{n(n+1)/2}n(2i)^n\frac{1}{r^{2n}}V(B_r),$$

where $V(B_r)$ is the Lebesgue volume of the ball of radius r in $\mathbf{R}^{2n}$, and is classically known to be equal to $\pi^n r^{2n}/n!$. Thus finally we see that our expression is equal to

$$f(0)(-1)^{n(n+1)/2}\frac{(2\pi i)^n}{(n-1)!}.$$

This proves Cauchy's theorem.

§6. The residue theorem

Let f be an analytic function in an open set U of $\mathbf{C}^n$. The set of zeros of f is called a **divisor**, which we denote by $V = V_f$. In the neighborhood of a regular point a, that is a point where $f(a) = 0$ but some complex partial derivative of f is not zero, the set V is a complex submanifold of U. In fact, if, say, $D_n f(a) \neq 0$, then the map

$$(z_1,\ldots,z_n) \mapsto (z_1,\ldots,z_{n-1},f(z))$$

gives a local analytic chart (analytic isomorphism) in a neighborhood of a. Thus we may use f as the last coordinate, and locally V is simply obtained by the projection on the set $f = 0$. This is a special case of the complex analytic inverse function theorem.

It is always true that the function $\log|f|$ is locally in $\mathscr{L}^1$. We give the proof only in the neighborhood of a regular point a. In this case, we can change f by a chart (which is known as a change-of-variable formula), and we may therefore assume that $f(z) = z_n$. Then $\log|f| = \log|z_n|$, and the Lebesgue integral decomposes into a simple product integral, which reduces our problem to the case of one variable, that is to the fact that $\log |z|$ is locally integrable near 0 in the ordinary complex plane. Writing $z = re^{i\theta}$, our assertion is obvious since the function $r \log r$ is locally integrable near 0 on the real line.

For the next theorem, it is convenient to let

$$d^c = -i(\partial - \bar{\partial}).$$

Note that

$$dd^c = 2i\, \partial\bar{\partial}.$$

The advantage of dealing with d and d^c is that they are real operators.

The next theorem, whose proof consists of repeated applications of Stokes' theorem, is due to Poincaré. It relates integration on V and U by a suitable kernel.

Residue theorem. *Let f be analytic on an open set U of $\mathbb{C}^n$ and let V be its divisor of zeros in U. Let ψ be a C^∞ form with compact support in U, of degree $2n - 2$ and type $(n - 1, n - 1)$. Then*

$$\int_V \psi = \frac{1}{2\pi} \int_U \log |f|\, dd^c\psi.$$

(As usual, the integral on the left is the integral of the restriction of ψ to V, and by definition, it is taken over the regular points of V.)

Proof. Since ψ and $dd^c\psi$ have compact support, the theorem is local (using partitions of unity). We give the proof only in the neighborhood of a regular point. Therefore we may assume that U is selected sufficiently small so that every point of the divisor of f in U is regular, and such that, for small ε, the set of points

$$U_\varepsilon = \{z \in U,\, |f(z)| \geq \varepsilon\}$$

is a submanifold with boundary in U. The boundary of U_ε is then the set of points z such that $|f(z)| = \varepsilon$. (Actually to make this set a submanifold we only need to select ε to be a regular value, which can be done for arbitrarily small ε by Sard's theorem.) For convenience we let S_ε be the boundary of U_ε, that is the set of points z such that $|f(z)| = \varepsilon$.

Since $\log |f|$ is locally in $\mathscr{L}^1$, it follows that

$$\int_{U_\varepsilon} \log |f|\, dd^c\psi = \lim_{\varepsilon \to 0} \int_{U_\varepsilon} \log |f|\, dd^c\psi.$$

Using the trivial identity

$$d(\log |f| \, d^c\psi) = d \log |f| \wedge d^c\psi + \log |f| \, dd^c\psi,$$

we conclude by Stokes' theorem that this limit is equal to

$$\lim_{\varepsilon \to 0} \left[\int_{S_\varepsilon} \log |f| \, d^c\psi - \int_{U_\varepsilon} d \log |f| \wedge d^c\psi \right].$$

The first integral under the limit sign approaches 0. Indeed, we may assume that $f(z) = z_n = re^{i\theta}$. On S_ε we have $|f(z)| = \varepsilon$, so $\log |f| = \log \varepsilon$. There exist forms ψ_1, ψ_2 in the first $n-1$ variables such that

$$d^c\psi = \psi_1 \wedge dz_n + \psi_2 \wedge d\bar z_n,$$

and the restriction of dz_n to S_ε is equal to

$$\varepsilon i e^{i\theta} \, d\theta,$$

with a similar expression for $d\bar z_n$. Hence our boundary integral is of type

$$\varepsilon \log \varepsilon \int_{S_\varepsilon} \omega,$$

where ω is a bounded form. From this it is clear that the limit is 0.

Now we compute the second integral. Since ψ is assumed to be of type $(n - 1, n - 1)$ it follows that for any function g,

$$\partial g \wedge \partial \psi = 0 \quad \text{and} \quad \bar\partial g \wedge \bar\partial \psi = 0.$$

Replacing d and d^c by their values in terms of ∂ and $\bar\partial$, it follows that

$$-\int_{U_\varepsilon} d \log |f| \wedge d^c\psi = \int_{U_\varepsilon} d^c \log |f| \wedge d\psi.$$

We have

$$d(d^c \log |f| \wedge \psi) = dd^c \log |f| \wedge \psi - d^c \log |f| \wedge d\psi.$$

Furthermore dd^c is a constant times $\partial\bar\partial$, and $dd^c \log |f| = 0$ in any open set where $f \neq 0$, because

$$\partial\bar\partial \log |f| = \tfrac{1}{2} \partial\bar\partial(\log f + \log \bar f) = 0$$

since $\partial \log \bar{f} = 0$ and $\bar{\partial} \log f = 0$ by the local analyticity of $\log f$. Hence we obtain the following values for the second integral by Stokes:

$$\int_{U_\varepsilon} d^c \log |f| \wedge d\psi = \int_{S_\varepsilon} d^c \log |f| \wedge \psi.$$

Since

$$d^c \log |f| = -\frac{i}{2} (\partial - \bar{\partial})(\log f + \log \bar{f})$$

$$= -\frac{i}{2} \left(\frac{dz_n}{z_n} - \frac{d\bar{z_n}}{\bar{z_n}} \right)$$

(always assuming $f(z) = z_n$), we conclude that if $z_n = re^{i\theta}$, then the restriction of $d^c \log |f|$ to S_ε is given by

$$\operatorname{res}_{S_\varepsilon} d^c \log f = d\theta.$$

Now write ψ in the form

$$\psi = \psi_1 + \psi_2$$

where ψ_1 contains only $dz_j, d\bar{z_j}$ for $j = 1, \ldots, n-1$ and ψ_2 contains dz_n or $d\bar{z_n}$. Then the restriction of ψ_2 to S_ε contains $d\theta$, and consequently

$$\int_{S_\varepsilon} d^c \log |f| \wedge \psi = \int_{S_\varepsilon} d\theta \wedge (\psi_1 \mid S_\varepsilon).$$

The integral over S_ε decomposes into a product integral, with respect to the first $n-1$ variables, and with respect to $d\theta$. Let

$$\int^{(n-1)} \psi_1(z) \mid S_\varepsilon = g(z_n).$$

Then simply by the continuity of g we get

$$\lim_{\varepsilon \to 0} \frac{1}{2\pi} \int_0^{2\pi} g(\varepsilon e^{i\theta}) \, d\theta = g(0).$$

Hence

$$\lim_{\varepsilon \to 0} \int_{S_\varepsilon} d\theta \wedge (\psi_1 \mid S_\varepsilon) = \int_{z_n = 0} \psi_1.$$

But the restriction of ψ_1 to the set $z_n = 0$ (which is precisely V) is the same as the restriction of ψ to V. This proves our theorem.

APPENDIX

The Spectral Theorem

The following is a set of notes from a seminar of Von Neumann around 1950.

§1. Hilbert space

Let $\mathbf{E}$ be a vector space over the complex. (The real theory follows exactly the same pattern.) By an **inner product** on $\mathbf{E}$ we mean a bilinear pairing $\langle x, y \rangle \in \mathbf{C}$ of $\mathbf{E} \times \mathbf{E}$ into $\mathbf{C}$ such that, for all complex numbers α, we have:

$$\langle \alpha x, y \rangle = \alpha \langle x, y \rangle, \qquad \langle x, y \rangle = \overline{\langle y, x \rangle}$$

$\langle x, x \rangle \geqq 0$ and equals 0 if and only if $x = 0$.

We have the **Schwartz inequality**:

$$|\langle x, y \rangle|^2 \leqq \langle x, x \rangle \langle y, y \rangle$$

whose proof is as follows. For all α, β complex,

$$0 \leqq \langle \alpha x + \beta y, \alpha x + \beta y \rangle = \alpha \bar{\alpha} \langle x, x \rangle + \beta \bar{\alpha} \langle x, y \rangle + \alpha \bar{\beta} \langle x, y \rangle + \beta \bar{\beta} \langle y, y \rangle.$$

We let $\alpha = \langle y, y \rangle$ and $\beta = -\langle x, y \rangle$. The inequality drops out.

We define the **norm** of a vector x to be $\langle x, x \rangle^{1/2}$ and denote it by $|x|$. Using the Schwartz inequality, one sees that $|x|$ defines a metric on $\mathbf{E}$, the distance between x and y being $|x - y|$. The norm is continuous.

We write $x \perp y$ and say that x is **perpendicular** to y if $\langle x, y \rangle = 0$.

The following identities are useful and trivially proved.

Parallelogram law: $|x + y|^2 + |x - y|^2 = 2|x|^2 + 2|y|^2$.

Pythagoras theorem: If $x \perp y$, then $|x + y|^2 = |x|^2 + |y|^2$.

A **Hilbert space** is an inner product space which is complete under the induced metric. For the rest of this appendix, a **subspace** will always mean a closed subspace, with its structure of Hilbert space induced by that of $\mathbf{E}$.

Lemma 1. *Let* **F** *be a subspace of* **E**, *let* $x \in$ **E**, *and let*

$$a = \inf |x - y|$$

the inf *taken over all* $y \in$ **F**. *Then there exists an element* $y_0 \in$ **F** *such that* $a = |x - y_0|$.

Proof. Let y_n be a sequence in **F** such that $|y_n - x|$ tends to a. We must show that y_n is Cauchy. By the parallelogram law,

$$|y_n - y_m|^2 = 2|y_n - x|^2 + 2|y_m - x|^2 - 4|\tfrac{1}{2}(y_n + y_m) - x|^2$$

$$\leqq 2|y_n - x|^2 + 2|y_m - x|^2 - 4a^2$$

which shows that y_n is Cauchy, converging to some vector y_0. The lemma follows by continuity.

Theorem 1. *If* **F** *is a subspace properly contained in* **E**, *then there exists a vector* z *in* **E** *which is perpendicular to* **F** *(and* $\neq 0$).

Proof. Let $x \in$ **E** and $x \notin$ **F**. Let y_0 be an element of **F** which is at minimal distance from x (use Lemma 1). Let a be this distance and let $z = y_0 - x$. After a translation, we may assume that $z = x$, so that $|x| = a$. For any complex number α and $y \in$ **F** we have $|x + \alpha y| \geqq a$, whence

$$\langle x + \alpha y, x + \alpha y \rangle = |x|^2 + \bar{\alpha}\langle x, y \rangle + \alpha\overline{\langle x, y \rangle} + \alpha\bar{\alpha}|y|^2$$

$$\geqq a^2.$$

Put $\alpha = t\overline{\langle x, y \rangle}$. We get a contradiction for small values of t.

§2. *Functionals and operators*

A linear map A from a Hilbert space **E** to a Hilbert space **H** is **bounded** if there exists a positive real number α such that

$$|Ax| \leqq \alpha|x|$$

for all $x \in$ **E**. The norm of A, denoted by $|A|$ is the inf of all such α.

Proposition 1. *A linear map is bounded if and only if it maps the unit sphere on a bounded subset, if and only if it is continuous.*

Proof. Clear.

A **functional** is a continuous linear map into **C**. Functionals are bounded. We have the fundamental:

Representation theorem. *A linear map* $\lambda: \mathbf{E} \to \mathbf{C}$ *is bounded if and only if there exists* $y \in \mathbf{E}$ *such that* $\lambda(x) = \langle x, y \rangle$ *for all* $x \in \mathbf{E}$. *If such a* y *exists, it is unique.*

Proof. If $\lambda(x) = \langle x, y \rangle$ then the Schwartz inequality shows that it is bounded, with bound $|y|$. It is obvious that y is unique.

Conversely, let λ be bounded. Let $\mathbf{F}$ be the kernel of λ. Then $\mathbf{F}$ is a subspace. If $\mathbf{E} = \mathbf{F}$ then everything is trivial. If $\mathbf{E} \neq \mathbf{F}$, then there exists $z \in \mathbf{F}$, $z \notin \mathbf{E}$ such that z is perpendicular to $\mathbf{F}$ by Theorem 1. We contend that some multiple $y = \alpha z$ does it. A necessary condition on α is that

$$\langle z, \alpha z \rangle = \bar{\alpha}|z|^2.$$

This is also sufficient. Namely, $x - (\lambda(x)|\lambda(z))z$ lies in $\mathbf{F}$. Put $\alpha = \overline{\lambda(z)}/|z|^2$. Then one sees at once that $\lambda(x) = \langle x, y \rangle$ as was to be shown.

By an **operator** we shall always mean a continuous linear map of a space into itself.

It is straightforward to show that operators form a Banach space, and in fact a normed ring. In other words, in addition to the Banach space property, we have

$$|AB| \leq |A|\,|B|.$$

Proposition 2. *If A is an operator and $\langle Ax, x \rangle = 0$ for all x, then $A = O$.*

Proof. This follows from the polarization identity,

$$\langle A(x+y), (x+y) \rangle - \langle A(x-y), (x-y) \rangle = 2[\langle Ax, y \rangle + \langle Ay, x \rangle].$$

Replace x by ix. Then we get

$$\langle Ax, y \rangle + \langle Ay, x \rangle = 0$$

$$i\langle Ax, y \rangle - i\langle Ay, x \rangle = 0$$

for all x, y whence $\langle Ax, y \rangle = 0$ and $A = O$.

The above proposition is valid only in the complex case.

In the real case, we shall need it only when A is symmetric (see below), in which case it is equally clear. A similar remark applies to the next result.

Lemma. *Let A be an operator, and c a number such that*

$$|\langle Ax, x \rangle| \leq c|x|^2$$

for all $x \in \mathbf{E}$. Then for all x, y we have

$$|\langle Ax, y \rangle| + |\langle x, Ay \rangle| \leq 2c|x|\,|y|.$$

Proof. By the polarization identity,

$$2|\langle Ax, y\rangle + \langle Ay, x\rangle| \leqq c|x + y|^2 + c|x - y|^2 = 2c(|x|^2 + |y|^2).$$

Hence

$$|\langle Ax, y\rangle + \langle Ay, x\rangle| \leqq c(|x|^2 + |y|^2).$$

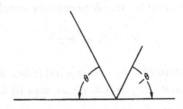

We multiply y by $e^{i\theta}$ and thus get on the left-hand side

$$|e^{-i\theta}\langle Ax, y\rangle + e^{i\theta}\langle Ay, x\rangle|.$$

The right-hand side remains unchanged, and for suitable θ, the left-hand side becomes

$$|\langle Ax, y\rangle| + |\langle Ay, x\rangle|.$$

(In other words, we are lining up two complex numbers by rotating one by θ and the other by $-\theta$.) Next we replace x by tx and y by y/t for t real and $t > 0$. Then the left-hand side remains unchanged, while the right-hand side becomes

$$g(t) = t^2|x|^2 + \frac{1}{t^2}|y|^2.$$

The point at which $g'(t) = 0$ is the unique minimum, and at this point t_0 we find that

$$g(t_0) = |x|\,|y|.$$

This proves our lemma.

In our applications, we need the lemma only when A is self-adjoint (i.e. symmetric, see below), in which case it is even more trivial.

For fixed y, the function of x is given by $\langle Ax, y\rangle$ is a functional (bounded because of the Schwartz inequality). Hence by the representation theorem, there exists an element y^* such that $\langle Ax, y\rangle = \langle x, y^*\rangle$ for all x. We define A^*, the **adjoint** of A, by letting $A^*y = y^*$. Since y^* is unique, we see that A^* is the unique operator such that

$$\langle Ax, y \rangle = \langle x, A^*y \rangle$$

for all x, y in $\mathbf{E}$.

Theorem 2. *We have:*

$$(A + B)^* = A^* + B^* \qquad A^{**} = A$$
$$(\alpha A)^* = \bar{\alpha} A^* \qquad |A^*| = |A|$$
$$(AB)^* = B^* A^* \qquad |AA^*| = |A|^2$$

and the mapping $A \to A^*$ *is continuous.*

Proof. Exercise for the reader.

§3. Hermitian operators

We shall say that an operator A is **symmetric** (or **hermitian**) if $A = A^*$.

Proposition 3. *A is hermitian if and only if* $\langle Ax, x \rangle$ *is real for all* x.

Proof. Let A be hermitian. Then $\overline{\langle Ax, x \rangle} = \langle x, Ax \rangle = \langle Ax, x \rangle$. Conversely, $\langle Ax, x \rangle = \overline{\langle Ax, x \rangle} = \langle x, Ax \rangle = \langle A^*x, x \rangle$ implies that

$$\langle (A - A^*)x, x \rangle = 0$$

whence $A = A^*$ by polarization.

Proposition 4. *Let A be a hermitian operator. Then* $|A|$ *is the greatest lower bound of all values c such that*

$$|\langle Ax, x \rangle| \leq c|x|^2$$

for all x, or equivalently, the sup of all values $|\langle Ax, x \rangle|$ *taken for x on the unit sphere in E.*

Proof. When A is hermitian we obtain

$$|\langle Ax, y \rangle| \leq c|x|\,|y|$$

for all $x, y \in E$, so that we get $|A| \leq c$ in the lemma of §2. On the other hand, $c = |A|$ is certainly a possible value for c by the Schwartz inequality. This proves our proposition.

Proposition 4 allows us to define an ordering in the space of hermitian operators. If A is hermitian, we define $A \geq O$ and say that A is **positive** if $\langle Ax, x \rangle \geq 0$ for all $x \in E$. If A, B are hermitian we define $A \geq B$ if $A - B \geq O$. This is indeed an ordering; the usual rules hold: If $A_1 \geq B_1$ and $A_2 \geq B_2$, then

$$A_1 + A_2 \geq B_1 + B_2.$$

If c is a real number ≥ 0 and $A \geq O$, then $cA \geq O$. So far, however, we say nothing about a product of positive hermitian operators AB, even if $AB = BA$. We shall deal with this question later.

Let c be a bound for A. Then $|\langle Ax, x \rangle| \leq c|x|^2$ and consequently

$$-cI \leq A \leq cI.$$

For simplicity, if α is real, we sometimes write $\alpha \leq A$ instead of $\alpha I \leq A$, and similarly we write $A \leq \beta$ instead of $A \leq \beta I$. If we let

$$\alpha = \inf_{|x|=1} \langle Ax, x \rangle \quad \text{and} \quad \beta = \sup_{|x|=1} \langle Ax, x \rangle,$$

then we have

$$\alpha \leq A \leq \beta,$$

and from Proposition 3,

$$|A| = \max(|\alpha|, |\beta|).$$

Let p be a polynomial with real coefficients, and let A be a hermitian operator. Write

$$p(t) = a_n t^n + \cdots + a_0.$$

We define

$$p(A) = a_n A^n + \cdots + a_0 I.$$

We let $\mathbf{R}[A]$ be the algebra generated over $\mathbf{R}$ by A, that is the algebra of all operators $p(A)$, where $p(t) \in \mathbf{R}[t]$. We wish to investigate the closure of $\mathbf{R}[A]$ in the (real) Banach space of all operators. We shall show how to represent this closure as a ring of continuous functions on some compact subset of the reals. First, we observe that the hermitian operators form a closed subspace of $L(\mathbf{E}, \mathbf{E})$, and that $\mathbf{R}[A]$ is a closed subspace of the space of hermitian operators.

We can find real numbers α, β such that

$$\alpha I \leq A \leq \beta I.$$

We shall prove that if p is a real polynomial which takes on positive values on the interval $[\alpha, \beta]$, then $p(A)$ is a positive operator.

The fundamental theorem is the following.

Theorem 3. *Let α, β be real and $\alpha I \leq A \leq \beta I$. Let p be a real polynomial, positive in the interval $\alpha \leq t \leq \beta$. Then $p(A)$ is a positive operator.*

Proof. We shall need the following obvious facts.

If A, B are Hermitian, A commutes with B, and $A \geq 0$, then AB^2 is positive.

If $p(t)$ is quadratic, of type $p(t) = t^2 + at + b$ and has imaginary roots, then

$$p(t) = \left(t + \frac{a}{2}\right)^2 + \left(b - \frac{a^2}{4}\right)$$

is a sum of squares.

A sum of squares times a sum of squares is a sum of squares (if they commute).

If $p(t)$ has a root γ in our interval, then the multiplicity of γ is even.

Our theorem now follows from the following purely algebraic statement.

Let $\alpha \leq t \leq \beta$ be a real interval, and $p(t)$ a real polynomial which is positive in this interval. Then $p(t)$ can be written:

$$p(t) = c\Big[\sum Q_i^2 + \sum (t - \alpha)Q_j^2 + \sum (\beta - t)Q_k^2\Big]$$

where Q_ν^2 just denotes the square of some polynomial and c is a number ≥ 0.

In order to prove this, we split $p(t)$ over the real numbers into linear and quadratic factors. If a root γ is $\leq \alpha$, then we write

$$(t - \gamma) = (t - \alpha) + (\alpha - \gamma)$$

and note that $(\alpha - \gamma)$ is a square. If a root γ is $\geq \beta$, then we write

$$(\gamma - t) = (\gamma - \beta) + (\beta - t)$$

with $(\gamma - \beta)$ a square. We can then write, after expanding out the factorization of $p(t)$,

$$p(t) = c\Big[\sum Q_i^2 + \sum (t - \alpha)Q_j^2 + \sum (\beta - t)Q_k^2 + \sum (t - \alpha)(\beta - t)Q_l^2\Big]$$

with some constant c and Q_ν^2 standing for the square of some polynomial. Note that c is positive since $p(t)$ is positive on the interval. Our last step reduces the bad last term to the preceding ones by means of the identity

$$(t - \alpha)(\beta - t) = \frac{(t - \alpha)^2(\beta - t) + (t - \alpha)(\beta - t)^2}{\beta - \alpha}.$$

Corollary 1. *If $a \leqq p(t) \leqq b$ in the interval, then*

$$aI \leqq p(A) \leqq bI.$$

Suppose that $\alpha I \leqq A \leqq \beta I$. If $p(t)$ is a real polynomial, we define as usual

$$\|p\| = \sup |p(t)|$$

with t ranging over the interval.

Corollary 2. *Let $\alpha I \leqq A \leqq \beta I$. Let $p(t)$ be a real polynomial. Then* $|p(A)| \leqq \|p\|$.

Proof. Let $q(t) = \|p\| \pm p(t)$. Then $q(t)$ is $\geqq 0$ on the interval. Hence $q(A) \geqq O$ and our assertion follows at once.

As usual, we consider the continuous functions on the interval as a Banach space. If f is any continuous function on the interval, then by the Weierstrass approximation theorem, we can find a sequence of polynomials $\{p_n\}$ approaching f uniformly on this interval. We define $f(A)$ as the limit of $p_n(A)$. From Corollary 2 we deduce that $\{p_n(A)\}$ is a Cauchy sequence, and that its limit does not depend on the choice of the sequence $\{p_n\}$. Furthermore, by continuity, our corollary generalizes to continuous functions, so that $|f(A)| \leqq \|f\|$.

We see that the map $f \mapsto f(A)$ is a continuous homomorphism from the Banach algebra of continuous functions on the interval into the closure of the subalgebra generated by A.

Proposition 5. *Let A be a positive operator. Then there exists an operator B in the closure of the algebra generated by A such that $B^2 = A$.*

Proof. The continuous function $t^{1/2}$ maps on $A^{1/2}$.

Corollary. *The product of two positive, commuting Hermitian operators is again positive.*

Proof. Let A, C be Hermitian and $AC = CA$. If B is as in Proposition 5, then

$$\langle ACx, x \rangle = \langle B^2Cx, x \rangle = \langle BCx, Bx \rangle = \langle CBx, Bx \rangle \geqq 0.$$

The kernel of our homomorphism from the continuous functions to the operators is a closed ideal. Its zeros form a closed set called the **spectrum** of A and denoted by $\sigma(A)$.

Lemma 2. *Let X be a compact set, R the ring of continuous functions on X, and $\mathfrak{a}$ a closed ideal of R, $\mathfrak{a} \neq R$. Let C be the closed set of zeros of $\mathfrak{a}$. Then C is not empty and if a function $f \in R$ vanishes on C, then $f \in \mathfrak{a}$.*

Proof. Given ε, let U be the open set where $|f| < \varepsilon$. Then $X - U$ is closed. For each point $t \in X - U$ there exists a function $g \in \mathfrak{a}$ such that $g(t) \neq 0$ in a neighborhood of t. These neighborhoods cover $X - U$, and so does a finite number of them, with functions $g_1, \ldots, g_r$. Let $g = g_1^2 + \cdots + g_r^2$. Then $g \in \mathfrak{a}$. Our function g has a minimum on $X - U$ and for n large, the function

$$f \frac{ng}{1 + ng}$$

is close to f on $X - U$ and is $< \varepsilon$ on U, which proves what we wanted.

We now redefine the norm of a continuous function f to be

$$\|f\|_A = \sup_{t \in \sigma(A)} |f(t)|.$$

Theorem 4. *The map*

$$f(t) \mapsto f(A)$$

induces a Banach-isomorphism (i.e. norm-preserving) of the Banach algebra of continuous functions on $\sigma(A)$ onto the closure of the algebra generated by A.

Proof. We have already proved that our map is an algebraic isomorphism and that $|f(A)| \leq \|f\|_A$. In order to get the reverse inequality, we shall prove:

If $f(A) \geq O$, then $f(t) \geq 0$ on the spectrum of A. Indeed, if $f(c) < 0$ for some $c \in \sigma(A)$, we let $g(t)$ be a function which is 0 outside a small neighborhood of c, is ≥ 0 everywhere, and is > 0 at c. Then $g(A)$ and $g(A)f(A)$ are both ≥ 0 (by the corollary of Proposition 5). But $-g(t)f(t) \geq 0$ gives $-g(A)f(A) \geq O$ whence $g(A)f(A) = O$. Since $g(t)f(t)$ is not 0 on the spectrum of A, we get a contradiction.

Let now $s = |f(A)|$. Then $sI - f(A) \geq O$ implies that $s - f(t) \geq 0$, which proves the theorem.

From now on, the norm on continuous functions will refer to the spectrum. All that remains to do is identify our spectrum with what can be called the **general spectrum**, that is those complex values ξ such that $A - \xi$ is not invertible. (By invertible, we mean having an inverse which is an operator.)

Theorem 5. *The general spectrum is compact, and in fact, if ξ is in it, then $|\xi| \leq |A|$. If A is Hermitian, then the general spectrum is equal to $\sigma(A)$.*

Proof. The complement of the general spectrum is open, because if $A - \xi_0$ is invertible, and ξ is close to ξ_0, then $(A - \xi_0)^{-1}(A - \xi)$ is close to I, hence invertible, and $A - \xi$ is also invertible. Furthermore, if $\xi > |A|$, then $|A/\xi| < 1$ and hence $I - (A/\xi)$ is invertible (by the power series argument). So is $A - \xi$ and we are done. Finally, suppose that ξ is in the general spectrum. Then ξ is real. Otherwise, let $g(t) = (t - \xi)(t - \bar\xi)$. Then $g(t) \neq 0$ on $\sigma(A)$ and $h(t) = 1/g(t)$ is its inverse. From this we see that $A - \xi$ is invertible.

Suppose ξ is not in the spectrum. Then $t - \xi$ is invertible and so is $A - \xi$.

Suppose ξ is in the spectrum. After a translation, we may suppose that 0 is in the spectrum. Consider the function $g(t)$ as follows:

$$g(t) = \begin{cases} 1/|t| & |t| \geq 1/N \\ N & |t| \leq 1/N \end{cases}$$

(g is positive and has a peak at 0.) If A is invertible, $BA = I$, then from $|tg(t)| \leq 1$ we get $|Ag(A)| \leq 1$ and hence $|g(A)| \leq |B|$. But $g(A)$ becomes arbitrarily large as we take N large. Contradiction.

Theorem 6. *Let S be a set of operators of the Hilbert space E, leaving no closed subspace invariant except 0 and E itself. Let A be a Hermitian operator such that $AB = BA$ for all $B \in S$. Then $A = \lambda I$ for some real number λ.*

Proof. It will suffice to prove that there is only one element in the spectrum of A. Suppose there are two, $\lambda_1 \neq \lambda_2$. There exist continuous functions f, g on the spectrum such that neither is 0 on the spectrum, but fg is 0 on the spectrum. For instance, one may take for f, g the functions whose graph is indicated on the next diagram.

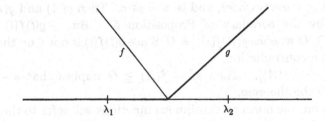

We have $f(A)B = Bf(A)$ for all $B \in S$ (because B commutes with real polynomials in A, hence with their limits). Hence $f(A)\mathbf{E}$ is invariant under S because

$$Bf(A)\mathbf{E} = f(A)B\mathbf{E} \subset f(A)\mathbf{E}.$$

Let $\mathbf{F}$ be the closure of $f(A)\mathbf{E}$. Then $\mathbf{F} \neq 0$ because $f(A) \neq O$. Furthermore, $\mathbf{F} \neq \mathbf{E}$ because $g(A)f(A)\mathbf{E} = 0$ and hence $g(A)\mathbf{F} = 0$. Since $\mathbf{F}$ is obviously invariant under S, we have a contradiction.

Corollary. *Let S be a set of operators of the Hilbert space* **E**, *leaving no closed subspace invariant except* 0 *and* **E** *itself. Let A be an operator such that* $AA^* = A^*A$, $AB = BA$ *and* $A^*B = BA^*$ *for all* $B \in S$. *Then* $A = \lambda I$ *for some complex number* λ.

Proof. Write $A = A_1 + iA_2$ where A_1, A_2 are Hermitian and commute (e.g. $A_1 = (A + A^*)/2$). Apply the theorem to each one of A_1 and A_2 to get the result.

Bibliography

[1] R. ABRAHAM, *Lectures of Smale on differential topology*, Columbia University, New York.

[2] W. AMBROSE, R. S. PALAIS, I. M. SINGER, "Sprays," *Acad. Brasileira de Ciencias*, **32** (1960) pp. 163–178.

[3] R. BOTT, "Morse Theory and its applications to homotopy theory," Lecture notes by A. Van de Ven, Bonn, 1960.

[4] N. BOURBAKI, *Espaces vectoriels topologiques*, Hermann, Paris,

[5] N. BOURBAKI, *Fascicule de résultats des variétés*, Hermann, Paris,

[6] N. BOURBAKI, *Topologie generale*, 2nd ed., Hermann, Paris, Chapter 9.

[7] D. G. EBIN and J. MARSDEN, "Groups of diffeomorphisms and the motion of an incompressible fluid," *Annals Math.*, **92** (1970) pp. 102–163.

[8] J. EELLS, Jr., "Alexander-Pontrjagin duality in function spaces," *Proc. Symposia in Pure Mathematics*, **3** Am. Math. Soc. (1961) pp. 109–129.

[9] J. EELLS, Jr., "On the geometry of function spaces," *Symposium de Topologia Algebrica*, Mexico (1958) pp. 303–307.

[10] J. EELLS, Jr., "On submanifolds of certain function spaces," *Proc. natn. Acad. Sci.* **45** (1959) pp. 1520–1522.

[11] M. C. IRWIN, "On the stable manifold theorem," *Bull. London math. Soc.* (1970) pp. 68–70.

[12] S. LANG, *Real Analysis*, Addison-Wesley, Reading, Mass., 1969.

[13] L. LOOMIS and S. STERNBERG, *Advanced Calculus*, Addison-Wesley, Reading, Mass., 1968.

[14] B. MAZUR, "Stable equivalence of differentiable manifolds," *Bull. Am. math. Soc.* **67** (1961) pp. 377–384.

[15] J. MILNOR, *Der Ring der Vectorraumbundel eines topologischen Raümes*, Bonn, 1959.

[16] J. MILNOR, *Differentiable Structures*, Princeton, 1961.

[17] J. MILNOR, *Differential Topology*, Princeton, 1958.

[18] J. MILNOR, "Morse Theory," *Ann. Math. Stud.*, **51** Princeton, 1963.

[19] J. MOSER, "A new technique for the construction of solutions for nonlinear differential equations," *Proc. natn. Acad. Sci.*, **47** (1961) pp. 1824–1831.

[20] J. MOSER, "On the volume element on a manifold," *Am. math. Soc. Trans.* **120** (1965) pp. 286–294.

[21] J. NASH, "The embedding problem for Riemannian manifolds," *Ann. Math.* **63** (1956) pp. 20–63.

[22] R. PALAIS, *Foundations of Global Analysis*, Benjamin, New York, 1968.

[23] R. PALAIS, "Morse theory on Hilbert manifolds," *Topology* (1963) pp. 299–340.

[24] Proceedings of the conference on global analysis, Berkeley, Calif., 1968; published by AMS, 1970.

[25] J. SCHWARTZ, "On Nash's implicit functional theorem," *Communs. Pure appl. Math.*, **13** (1960) pp. 509–530.

[26] S. SMALE, "Differentiable Dynamical systems," *Bull. Am. math. Soc.* (1967) pp. 747–817.

[27] S. SMALE, "Generalized Poincaré's conjecture in dimensions greater than four," *Ann. Math.* **74** (1961) pp. 391–406.

[28] S. SMALE, "Morse theory and a non-linear generalization of the Dirichlet problem," *Ann. Math.* (1964) pp. 382–396.

[29] S. STERNBERG, *Lectures on Differential Geometry*, Prentice-Hall, Englewood Cliffs, N.J., 1964.

Index

Printed in the United States
By Bookmasters

Printed in the United States
By Bookmasters